BRITAIN'S JET AGE

FROM THE JAVELIN TO THE VC10

Guy Ellis

Acknowledgements, copyright and an apology

I would like to thank everyone who has contributed to both volumes of *Britain's Jet Age*, both through their images and recollections and my editors Ken McGillivray and John Linnegar of http://www.editandtrain.com for their invaluable assistance.

Copyright is a very important concept in which I firmly believe and therefore make every effort to establish true ownership. I apologise for any errors which are completely unintentional.

The fabulous cover image on Volume 1 was incorrectly attributed by me and should have acknowledged the fine work of well-known photographer Richard Paver. My sincere apologies to Mr Paver for this mistake.

Front cover: Avro Vulcan XH558 and De Havilland DH-110 FAW2 Sea Vixen in formation at the RNAS Yeovilton Air Day, 9 July 2011. (Glen Pardoe)

Back cover: English Electric Canberra PR9 XH134 (G-OMHD) and Hawker Hunter T7 XL577 (G-XMHD) from the MidAir Squadron displaying at RAF Cosford in 2014. (Tony Hisgett)

First published 2016

Amberley Publishing
The Hill, Stroud
Gloucestershire, GL5 4EP

www.amberley-books.com

Copyright © Guy Ellis, 2016

The right of Guy Ellis to be identified as the Author of this work has been asserted in accordance with the Copyrights, Designs and Patents Act 1988.

ISBN 978 1 4456 4918 4 (print)
ISBN 978 1 4456 4919 1 (ebook)

British Library Cataloguing in Publication Data. A catalogue record for this book is available from the British Library.

Typeset in 10pt on 13pt Celeste.
Typesetting by Amberley Publishing.
Printed in the UK.

Jet Age – The Second Generation

The period from the mid-1950s to the early 1960s can be broadly classified as the era of second-generation jet aircraft. The aircraft had moved on from simply being jet-powered, advanced-piston designs to incorporate what had been learned during the Korean conflict and to take advantage of new technologies. Typically, the second generation was fitted with delta or swept wings and was powered by engines with afterburners, which provided sustained supersonic capability. Standard equipment comprised advanced on-board radar systems and guided air-to-air missiles. Only in the next generation would aerodynamics and avionics see further improvement. Most importantly, combat pilots of third-generation military jets no longer had to acquire their target visually.

First in Service

Gloster Javelin

The Javelin FAW7 was the second generation of the 'Flat Iron' fighter, incorporating many advanced features. Fitted with two powerful 11,000-lb thrust Sapphire SA7 engines, it was the first production Javelin to be armed with four Firestreak missiles. Its fuselage was lengthened and redesigned to reduce drag. Flight controls were considerably upgraded with the innovative Gloster-developed 'pitch auto stabilisation system, full hydraulically powered rudder operation with yaw stabilisation and electro-hydraulic three axes control autopilot with automatic approach and altitude control'. SA7R engines improved the subsequent FAW8 version, providing a reheat system that provided a 12 per cent increase in thrust above 10,000 feet. The later FAW9 introduced further developments: slightly drooped wing outer leading edges, liquid rather than cartridge starts and new flying controls. More than 80 per cent of the FAW7s built were converted to FAW9 standards and, of those, forty-four were fitted with non-retractable refuelling probes and designated FAW9R. Javelin production ended with 435 units produced, of which 133 had been built by Armstrong-Whitworth at Baginton, Coventry.

Conflict but no combat

The 1961 proposal to create a Malaysian Federation comprising Malaya and the British colonies of Borneo and Singapore was opposed by Indonesian President Sukarno, as he felt that the federation would allow Britain to deny Borneo real independence.

From April 1963 Indonesia supported guerrilla raids into Borneo in an undeclared war between Britain, Malaysia and Indonesia – the 'Confrontation'. Number 60 Squadron Javelin FAW9s were sent from RAF Tengah Singapore to Butterworth in Borneo after Indonesian landings on the Malaysian peninsula. In August 1964 further troop and aircraft reinforcements were despatched to Singapore, the squadron supported from December by FAW9R Javelins of No. 64 Squadron.

The FAW9R aircraft with the greater range were detached to Kuching and Labuan. Although designed for high level, the Javelins performed well on low-level patrols designed to provide early warnings of and to deter Indonesian Air Force (IAF) activity. An unknown in this mix of jet and piston engines was whether the Firestreak missiles would lock on to the piston engines of the IAF Mustang fighters.

At higher altitude, on occasion Indonesian bombers and transports were intercepted. Once, a Javelin shadowed a Tupolev Tu-16 'Badger' and noticed it was being tracked by the Badger's guns; the Javelin flashed its underneath to reveal four Firestreak missiles and the tracking ceased.

In addition to the low-level patrols, the crews performed quick reaction alert (QRA) duties and escorted resupply transport and helicopter flights. Ian Gibbon remembers,

> the impressive ground runs and having to do electrical checks to make sure the reheat system worked properly. You got a real feeling of the power as the undercarriage [was] compressed at full throttle especially as you had your head stuck in an access panel trying to use a test lamp to make sure the relays were changing correctly. [Ellis]

Although Singapore separated from Malaysia in 1965, Indonesia's Confrontation tactics failed and they were forced to accept the new Malaysian state on 11 August 1966. English Electric Javelin operations ended and the remaining two RAF squadrons flying them were based at RAF Tengah until they were disbanded in June 1967.

Javelins were employed in an ever-changing empire: in December 1965, a detachment of nine Gloster Javelin FAW9R aircraft from No. 29 Squadron was deployed from Akrotiri to Ndola in Zambia. Fully loaded with two 1,046-litre drop tanks on each wing and two bosom (or 'Sabrinas') 1,137-litre under-fuselage auxiliary fuel tanks, they were despatched to provide air defence for Zambia following the Rhodesian government's announcement of a unilateral declaration of independence (UDI) on 11 November. Within days of arriving in Ndola, four Javelins were relocated to Lusaka; two were kept on immediate standby, one at 10 minutes and the other 30 minutes. Neville Ward, RAF ground crew, recollected that,

> the only actual work we carried out was basic maintenance and once a week we dry cycled the engines, and then replaced all of the [starting] blanks. The IPN [Isopropyl nitrate], used in the cartridges to start the engines, was replaced fairly regularly due to evaporation. [Ellis]

Throughout their deployment until September 1966, few intercepts were recorded, although QRA duties kept them busy. A single Javelin XH897 remained in service with the Aircraft and Armament Experimental Establishment at Boscombe Down until 24 January 1975.

De Havilland Sea Vixen

De Havilland engineers explored radical improvements to the basic Sea Vixen design, but in the early 1960s they developed only a limited set:

> The difference in the two versions was immediately recognisable by protrusions in front of the leading edge of each wing. These ugly looking extensions, which were forward of where each tail boom joined the wing, carried extra fuel tanks. Although this increased the flying range without the need for drop tanks, they restricted the pilot's left and right field of vision. [Wrigley]

More significant was the Vixen's missile armament: armed with the Firestreak missile, it could also carry the Red Top air-to-air missile, four SNEB rocket pods and the air-to-ground Bullpup missile. Electronic counter measures (ECM) were installed to match the changes in aerial combat.

The first prototype FAW2 was XN684, flown on 1 June 1962 by Chris Capper, followed by the second conversion XN685. Both were used to run Red Top trails. XN684 made the last Vixen carrier landing on HMS *Eagle* before the vessel was decommissioned in 1972. XN685 is now on display at Midland Air Museum.

Production aircraft were delivered from March 1963, while the majority of FAW1s were converted to FAW2 standard. Number 899 Squadron were the first to embark the new mark aboard HMS *Eagle* in December 1964:

Marshallers synchronised the movement of aircraft to the launch pad by communicating through their radio headsets and, impatiently stamping their feet, they urgently directed aircraft ... the marshallers had absolute authority ...

Aircraft were manoeuvred onto one of the two steam catapult launch pads where they briskly taxied up against raised, hydraulically operated chocks. Once in position, the thick, cabled, metal bridle was swiftly looped round the steam-operated shuttle and attached to horns below each wing. The holdback bar ... was simultaneously raised from the deck to the aircraft rear attachment point. The chocks were then retracted and the blast deflector was raised.

In obedience to the launch controller's hand signals, the pilot lowered and locked the wings in the extended position, set the flaps and opened the throttles to take-off thrust. When ... ready for the launch, the pilot gave a sharp nod before pushing his head back against the headrest. ... After a moments' delay, the pilot was pushed back into his seat and, with a whooshing blast of steam, the aircraft was rapidly accelerated from a standstill to flying speed in just sixty metres. [Wrigley]

This was the age of the Fleet Air Arm (FAA) and the Royal Navy's carrier force. In 1961 HMS *Victorious* with its Vixens supported a Royal Marines landing in Kuwait to prevent an Iraqi invasion. In a similar operation, Vixens of HMS *Centaur* flew in support of marines landed in Tanganyika to reinforce government troops and protect incoming RAF transports.

The infamous 1966 Defence White Paper heralded the end of this era, but not before the carrier force was required to prevent oil reaching post-UDI Rhodesia via Mozambique and assisted in an operation overseeing the final withdrawal of British forces from Aden.

The aircraft was robustly constructed for carrier operations but,

from the time of their inauguration there were heavy aircrew losses ... From 1962 to 1970, thirty accidents accounted for the loss of fifty-one air crew members.

I enjoyed the Sea Vixen's performance capabilities. This was enhanced by the large area of its wings and control surfaces. I was especially impressed to find it was possible to maintain a 45 degree banked turn, and Mach 0.85 at 45 000 feet. [Wrigley]

This powerful sub-supersonic interceptor was at first hampered by the limited early missile technology and radar, which had no 'look-down' capability – by today's standards it was rather rudimentary.

In ground attack, the

> observer played an important and comforting role during steep rocket attacks over water. It was
> a great advantage to have the observer call the height above the water, which allowed the pilot to
> devote his attention to the gun-sight and to maintain the correct speed. [Wrigley]

The Sea Vixens' withdrawal from 1968 and their replacement by the McDonnell Douglas F-4
Phantom II marked the end of De Havilland's jet fighter designs. Out-surviving the Gloster
Javelin, the airframes had many years of service life left but rapid technological advancements
rendered their weapon systems obsolete. The powerful American McDonnell Phantom
replacement would ring the changes for the FAA.

Some surviving airframes were converted to U3 drones by Flight Refuelling at Tarrant
Rushton and a small number were used as target tugs known as TT2s. A few of these aircraft
flew until January 1974.

English Electric Canberra

English Electric had built their last aircraft in 1926 but continued to manufacture trolley
busses, trams and locomotives. However, during the Second World War, they began to build
Hampdens and Halifaxes on behalf of Handley Page, their production throughput so good that
the company decided to continue building aircraft after 1945.

Following his departure from Westlands, William 'Teddy' Petter was directed by the Air
Ministry to take his idea of a Mosquito replacement to English Electric. The original design
to meet the 1944 Ministry of Supply requirement for a high-speed, high-altitude bomber
was powered by a single engine; however, Rolls-Royce had made such progress with its
second-generation, axial-flow engines that the design was revised to house twin Avon engines
in the wing roots. Petter believed that the limited gain of a swept wing did not outweigh the
structural requirements to achieve 500 mph, so he went instead for a thin chord wing, with
engines mounted in the wing.

Four B3/45 A1 prototypes were ordered. However, development did not go smoothly, as the
new axial Avon RA1 engine was subject to power surges, and so the second prototype was
powered by the proven centrifugal Nene engines. In addition, the planned new radar bombing
system was larger and more complex than could be fitted in the A1, so it was replaced by a
visual bombsight and uprated wartime navigation systems. This entailed adding a third crew
member to the four subsequent B5/47 prototypes.

The first VN799 aircraft took off from Warton on 13 May 1949 with chief test pilot Roland
Beamont at the controls:

> It flew as if it had been flying for a hundred hours – it came off the ground at exactly the predicted
> speed with exactly the predicted stick forces, it was controllable with the fingertips, no muscle
> force and right from the first take-off, it was a delightful flying experience. [Beamont]

Partly in order to encourage Australian purchases, the chairman of English Electric suggested the name 'Canberra'. The first production aircraft WD929, a B Mk2, was used for the official ceremony when Australian Prime Minister, Robert Menzies, named the design 'Canberra' on 19 January 1951. WD929 was destroyed by a missile in 1959 after conversion to a drone.

At the outbreak of the Korean War, the Ministry of Supply ordered an increase in production and a number of other aircraft manufacturers were contracted to build the Canberra.

Records

Throughout the Jet Age records were broken, none more so than by the Canberra, which achieved nineteen flight and three altitude records. In 1951 it set the cross-Atlantic record at an average speed of 481.12 mph, bettered in 1952 to 605.52 mph. That same year a B5 VX185 criss-crossed the Atlantic twice in a day in a little over 10 hours. A Canberra PR 7 WT528 was flown from Tokyo to Kent over the North Pole on 25 May 1957, covering 51,100 nautical miles in 17 hours and 42 minutes. A Bristol Olympus-powered B2 WD952 set an altitude record of 65,890 feet in August 1955.

Marks

The airframe was versatile and durable, with more than twenty versions of the Canberra being developed. Identified as having the range, speed, altitude and payload flexibility to serve in a photo reconnaissance role, the B2 was modified, with a fuselage lengthened by 14 inches to house a forward camera bay. The PR3 carried one F49 vertical camera, six F52 oblique cameras and a crew of two.

Typical of the early Jet Age was the dearth of dual-seat trainers. Only in 1952 was the T4 crew trainer developed, with the instructor and student crew on ejection seats under the bubble canopy, and the navigator in the usual place behind and below the flight deck.

Engine development continued and by 1954 more powerful Avon 109 engines were fitted to the Canberra B Mk6 three-seater tactical bomber. In line with this, the PR3 was upgraded with the new power units and delivered to the RAF in 1954 as the PR Mk7. The Mk6 and Mk7 had a higher service ceiling than the first-generation Canberras, carried more fuel and could handle a heavier payload.

A new twin-seat night interdictor and tactical bomber, the B(I) Mk8, fitted to carry a nuclear weapon or a 20-mm gun pack and wing-mounted bombs, was built with a single offset canopy for the pilot, with the navigator radar operator in the usual position deep in the fuselage.

The apogee of the Canberra was the PR9, which itself led to an even more radical upgrade. Reconnaissance missions beyond the expected 60,000 feet range of ground-to-air missiles were required. The Americans had developed the Lockheed U2 and their high-altitude derivative of the Canberra, the RB-57D, while the PR9 was created as the RAF equivalent. It was powered by Avon 206 engines, developing double the thrust of the early units, to reach the high altitudes and fitted with powered controls to help pilots maintain control in the thin air at extreme heights.

The prototype PR9 first flew on 27 July 1958 and was followed by twenty-three production aircraft built by Shorts at Belfast. First entering service in 1960 with No. 58 Squadron at Wyton, it carried specialised photographic equipment designed to counter the degradation of image clarity typical of high-altitude photographs. It could carry a wider variety of imaging sensors than the U2. The PR9s were used both for military work and for significant civilian and joint military survey and mapping programmes.

The rapid development of jets required many pilots and navigators to clock up additional flying hours to hone their skills: the Canberra and Hunter second-generation jets achieved this cost-effectively. These older aircraft were less complex, offering high serviceability and utilisation. The PR9 was one of the most versatile aircraft in an air force's inventory: it had a range of 2,000 nautical miles, operated at all heights, and carried a variety of armaments and equipment.

Entering service only fourteen years after the invention of the jet engine, the Canberra became the longest-serving frontline aircraft when, after fifty-five years, the last PR9 landed at Kemble on 31 July 2006.

Foreign exports

The Canberra was a great export success. The Australians produced Canberras under licence at the Government Aircraft Factory (GAF) in Melbourne. The Americans also produced them under licence as the B57, and twenty-three other countries were supplied with aircraft that had often been retired, refurbished and sold on.

In 1953 Venezuela became the first export customer. The Fuerza Aerea Venezolana (FAV) eventually operated sixty aircraft – new builds, refurbished RAF stock and rebuilt aircraft covering seven different versions. Undergoing upgrades to keep pace with current avionics and weaponry, the fleet flew on into the 1990s.

In 1954 Ecuador ordered six Canberra bombers and twelve Gloster Meteor fighters, making the country the second South American nation to boast a tactical air force.

Peru followed with an order for eight new B(1)8 bombers in 1956, and then continued to add second-hand aircraft throughout the next four decades. The last additions in 1991 were a single T4 and five B(1)12 ex-South African Air Force (SAAF) aircraft.

Ten years later the Fuerza Aerea Argentina (FAA) purchased ten ex-RAF B2s and two T4s (known as B62 and B64s) – one BMk62 was still in service in 1999. The Canberras fought against their previous owners in the 1982 Falklands conflict and two were lost, one to a Harrier, the other shot down by HMS *Exeter*.

In the East the Canberra was also popular. Forty-eight were built under licence by the Australian GAF, with the first, A84-201, flying on 29 May 1953. Based on the B2, these aircraft were modified to meet local requirements and designated Mk20s. The Royal Australian Air Force (RAAF) supported the British forces in the Malaysian Confrontation in 1958. From 1967 until May 1971 the Canberra was used to support Australian and US ground forces in Vietnam. As the only aircraft in the conflict that could bomb visually, it earned a reputation as being extremely accurate at low level and, surprisingly, in the close support of ground troops. Final withdrawal from RAAF service took place in June 1982.

In 1958 New Zealand signed a contract for eleven aircraft, nine B(I)12s and two T13s. The main difference between these marks and the RAF examples were a different autopilot and an extra fuel in the bomb bay of the trainer. The Royal New Zealand Air Force (RNZAF) also supported the RAF during the Malaysian Confrontation, being based first in Singapore and then in Butterworth. The bombers were sold to India in 1970.

The Indian Air Force (IAF) became the largest operator of the Canberra following its selection in January 1957. Over the years the IAF operated seventy-two aircraft across eight variants. No. 5 Squadron was the first to be equipped with the B(I)58 and No. 106 Squadron reconnaissance specialists flew the first of the PR variants.

War was not foreign to these aircraft. In 1961 they participated in the Indian annexation of Portuguese India (Goa) and the war in the Congo and were put on standby in 1962 during a confrontation between India and China. They were highly utilised during the Indo-Pakistani wars, carrying out reconnaissance and bombing raids, notably on the Karachi oil installations. As late as 1999 they were used in a PR role during the Indo-Pakistan conflict over the Kargil District in Kashmir and were finally retired only in 2007.

In 1959 the Rhodesian Air Force received fifteen B2 bombers and three T4 trainers (all refurbished ex-RAF aircraft). Used extensively in the bush war of the 1970s and 1980s, many were retired due to fatigue; some were lost in action; a few served until 1979; and two hybrid aircraft were built of retired units. Neighbouring South Africa ordered six B(1)12s and three T4s from refurbished RAF stock in 1963. Also used extensively in southern African conflicts, all but one B(1)12 survived to be sold onto Peru. Two T4s are still on display.

At the outbreak of the Korean War, the United States Air Force (USAF) had only Second World War medium bombers. After competition between a series of manufacturers, the Canberra was selected for licensed production. The Glen L. Martin Company went on to design and build new developments, creating multiple variants across the 403 units produced between 1953 and 1957.

The American B57 exploited work done by English Electric in an earlier design exercise to improve the Canberra. Roland 'Bee' Beamont remembers:

We called it the P4 project. By then, however, the Ministry had already paid for the basic design ... and they weren't going to pay again. It was as simple as that. ... the Americans were able to take a fresh look and they incorporated a number of minor changes and, of course, the major one of the tandem two-seat cockpit. [Beamont]

It was the last tactical bomber in the USAF; fitted with powerful radars and laser-guided armaments, it could find targets in complete darkness and it was the first aircraft to launch a 'smart' or precision-guided bomb. In support of this role, B57s served in a tactical reconnaissance role, reporting back on ground attacks and troop movements. Strategically, it was used to photograph enemy cities and factories for planning raids by B52 strategic bombers. In less combative roles it performed weather-monitoring services, towed targets, was used for extensive tests of aircraft components and procedures and equipped the Air National Guard.

Still flying

To meet the requirement for a high-altitude reconnaissance aircraft, the RB57D had a lengthened wingspan – almost twice as long as that of the original aircraft – with an increased wing chord and more powerful engines, allowing it to fly at 60,000 feet. First delivered in 1956, they served for five years, when some airframes were upgraded to the RB57F and used to monitor the communist bloc's nuclear experiments. Re-designated WB57fs in 1968, they were phased out in the 1970s.

NASA operates three modified WB57s. These aircraft have always had the unique capability of ideal ceiling, range, payload and crew composition. Although it operates in the realm of the U-2 spyplane, the Canberra can carry four times the payload. All three aircraft have recently had upgraded autopilot systems, ejector seats and state-of-the-art communications to complement their enhanced data-gathering abilities. For a long time NASA flew only two of these aircraft but, in 2013, it selected aircraft 63-13295, which had been retired in 1972, and rebuilt it as NASA 927.

Built in 1957, Javelin FAW7 XH712 was used for Firestreak/Red Top trials until converted to an FAW9 version in 1959. It served with No. 23 Squadron, then No. 29 Squadron in Zambia, and was scrapped in 1968. (Jet Museum)

Javelin FAW7 XH768 is now preserved as 'XH707' at the Aviazione aviation museum near Rimini, Italy. It was converted to an FAW9 and served with Nos 11, 25 and 29 Squadrons before it was assigned to Cranwell for ground instruction in 1967, no doubt ensuring its survival. (Jet Age Museum)

Javelin FAW7 XH961 was delivered in 1959, and was converted a year later to FAW9 standard. Later, fitted with a refuelling probe, it became an FAW9R. Seen here with No. 60 Squadron undergoing maintenance at RAF Kuching in 1963/64, it was written off in a hard landing at RAF Tengah, Singapore. (Ian Gibbons)

Javelin FAW9R XH707/P of No. 60 Squadron over Sarawak, 1964. (Mark Taylor)

No. 29 Squadron FAW9R XH873/A taxiing at Ndola Airport, Zambia, in 1966. (Mervyn Blumberg)

Preparing XH889 for its next patrol over the Zambian border. (John Lewer, www.jetagemuseum.org)

FAW8 XH966, delivered in 1960 for testing at RAF AAEE Boscombe Down and then to No. 41 Squadron. It was scrapped in 1964. (BAE SYSTEMS)

Built in 1958, this FAW7 XH912 served with No. 25 Squadron before being converted in 1960 to an FAW9. It is in service here with No. 2 Squadron, before its final service with No. 228 OCU and scrapping in 1968. (True second-generation of Hunters and Lightnings in the background.) (BAE SYSTEMS)

Javelin FAW8 XJ122 and FAW8 served with No. 85 Squadron for only two years before being put in storage and then scrapped in 1965. (BAE SYSTEMS)

Gloster Javelin FAW9 XH844 at Farnborough in 1960. It caught fire during a starting sequence in 1962. (John Kendall)

The Sea Vixens of the HQ Squadron Yeovilton, 899NAS, spent short periods at sea. As this Vixen FAW1, marked 485VL, becomes airborne, the launching strip can be seen falling away underneath on HMS *Eagle;* No. 899 was the first squadron to evaluate and operate Sea Vixen FAW2 aircraft. (BAE SYSTEMS)

The Vixen prototype XF828, with everything hanging out, landing aboard an unidentified carrier. (BAE SYSTEMS)

Sea Vixen FAW1 XJ488 first flew in 1959 and spent its entire career as a test airframe until it was scrapped in 1972. Here it is seen undertaking in-flight refuelling with Sea Vixen FAW1 XJ521. This airframe was used as a ground trainer then fire trainer, and now the cockpit is preserved at Robertsbridge Aviation Society and Museum, Sussex. (BAE SYSTEMS)

XJ488 was assigned to ground and then fire-training duties. This is a view of the 'coal hole' where its cockpit is now preserved at Robertsbridge Aviation Society and Museum in Sussex. (Robertsbridge Aviation Society and Museum)

Vixens of No. 892 Squadron aboard HMS *Victorious* some time between October 1959 and February 1962, and an unidentified Scimitar, bearing the brunt of the heavy weather. (BAE SYSTEMS)

HMS *Ark Royal* in 1963: two Vixens, a Scimitar and the aeroplane guard Westland Whirlwind helicopter on the aft deck. (Andrew Patterson)

Sea Vixen FAW1 XN689, seen here aboard HMS *Eagle* in 1963 after its delivery to No. 890 NAS as 253R. After an upgrade to FAW2 in 1966, it was scrapped in 1969. (Andrew Patterson)

The high cost of conversion prevented more than about five Sea Vixens being modified as target drones – using them to teach pilots how to control drones remotely made more financial sense than to have such expensive aircraft destroyed by a missile. (Rob Finch)

De Havilland DH.110 Sea Vixen FAW2 XP954 experienced a double engine failure while flying for FRADU in November 1973; both crew ejected safely. (George Woods)

One D3 XP924/G-CVIX survived and was privately owned and operated by De Havilland Aviation Ltd at Bournemouth Aviation Museum, Bournemouth Airport, until it was moved to the Fly Navy Heritage Trust, to which it was donated in 2014. Now operated from Yeovilton by Naval Aviation Ltd, a subsidiary of Fly Navy Heritage Trust. (Rob Finch)

The English Electric factory in January 1951 with the first three production Canberra B2s, WD929 (second in line; it was used in the naming ceremony at Biggin Hill in 1952), WD930 and WD931, together with the third prototype VN828. In the background is an Avro Lincoln B2 of No. 101 Squadron, which operated the Canberra from 1951. (BAE SYSTEMS)

Seen here during a routine sortie, aircraft WJ874 was painted to represent the first prototype Canberra, VN799, on its first flight from Warton airfield, where it had been built in 1959. In all, 1,347 Canberras were produced, 773 of the 925 built in the UK having been destined for the RAF. (SAC A. K. Benson/MOD Crown, 2006)

Combat veteran WH646 Canberra T17A served with the Royal New Zealand Air Force during the confrontation with Indonesia. Built as a B2 in 1952, seen here in July 1987 at Fairford while serving with ECM No. 360 Squadron. (Kevin Slade)

Canberra WJ614 was built as a B2 in 1954, serving until 1972. Two years later it was converted to a TT18 target tug for the Royal Navy, serving until 1993. (Kevin Slade)

The August 1957 flight by Mike Randrupp, with Walter Shirley acting as flight observer, is commemorated on the side of WK163. The aircraft has been restored to B2 status and colour scheme. (Guy Ellis)

Three PR9 aircraft – XH131, XH134 and XH135 – ended Canberra frontline service when they touched down at Kemble in July 2006. Here XH135 draws attention prior to a fast taxi run at the 2007 Kemble air show. (Guy Ellis)

XH134 was returned to airworthy status and is seen here in 2014 at the Waddington air show. (Jonathan Francis)

Canberra B(I)58 IF922 was converted into a target tug role and operated by No. 1 Target Tug Unit, based at Puna Air Base, in 1990. (Bharat-Rakshak.com)

Indian Air Force Canberra PR57 IP990, operated by No. 106 Squadron from 1957, seen here at Yelahanka during Aero India 2005. Having survived a missile hit at Kargil, it was lost during a single-engine approach at Agra in December 2005. (Bharat-Rakshak.com)

Three ex-RAF Canberras – WK130, WK137 and WK138 (pictured here) – were purchased by the German government for trials and experiments. They were delivered to Erprobungstelle 61 and based at Oberpfaffenhofen, near Munich, in 1961. They were used for high-altitude calibrations, and two were fitted with a bomb-bay package of powerful cameras for survey and oblique surveillance. (Andy Chetwyn)

Peruvian Air force 242 ex-WH880, delivered to the RAF in 1953 and re-purchased by BAC in 1966, is seen here being test flown as G27-99 before delivery to Peru. It crashed in 1972. (BAE SYSTEMS)

WJ713/B-102 taking off at the Farnborough Air Show wearing UK registration G-AYHP and fitted with rocket pods for its COIN role. Built in 1953, it was sold to BAC and converted to B62 specifications for the Argentine Air Force as B-102. From 1970 it served with II Brigada Aerea, Grupo de Bombardeo 1, at Parana (Entre Rios) and took part in the Falklands War before being retired in 1998. Now preserved at Olivia City, Cordoba, Argentina, sporting a map of the Falklands and five mission markings. (Malcolm Nason)

The Chilean Air Force received three Canberra PR-9s in 1982: 341, 342 and 343. Operated by Grupo de Aviación No. 2 – a strategic reconnaissance unit – 343 was built for the RAF in 1960 as XH173. Sold to Chile in 1982, it was retired in 1995, finally being restored for display at the Chilean National Air and Space Museum at Los Cerrillos airport. (Museo Nacional Aeronautico y del Espacio Chile)

One of two T4s supplied to the South African Air Force in 1964. Serving with the RAF as WJ617 and originally built as a B2 in 1954, it was converted to a T4 in 1956. After 1990, 459 was retired and is now on display at Air Force Base Waterkloof, Pretoria. (Steve McLean)

B-57A (52-1419) was operated by the US National Oceanic and Atmospheric Administration (NOAA) out of Miami International Airport. With a civilian registration of N1005, it was fitted with a large nose radome and used to track hurricanes. Sold to a private flight-training company and eventually scrapped in 1988. (NOAA)

Built in 1963 by General Dynamics as an RB57F and subsequently much modified, N926NA is one of the three that flies with NASA. (NASA)

The Trainers

Folland Gnat

'Teddy' Petter, of Canberra fame, became known for his staunch faith in his light fighter idea, even though heavily engaged with the design of the English Electric P1, forerunner of the Lightning. He was in fact so enamoured with his idea that he left and joined Follands where he saw an opportunity to pursue his dream.

Derived from the even-more-diminutive Fo 139 Midge prototype, the Folland Gnat was offered in three roles – interceptor, close-support and carrier-borne fighter – and could achieve Mach 1.3 in a dive. Originally intended to be powered by the Bristol Saturn, when this was cancelled Follands used the modified Armstrong Siddeley Viper units until the new, more stable Orpheus engine became available from the Bristol Aeroplane Co. Ltd in November 1953.

The tiny fighter was fitted with: a lightweight, fully automatic ejector seat; gyro gunsight; VHF radio; limited radar; and a parachute to slow landing. The cockpit was compact, well laid out and its flying controls were light and precise. It was highly manoeuvrable and was notable for its inboard ailerons, which, when dropped, provided increased lift at a low speed, reducing speed. This made landing safer, while the strengthened undercarriage could be partially extended in flight and acted as a very effective air brake.

The Air Ministry and RAF officials did not support the Gnat, as their thinking was that a light fighter was not as versatile as a larger fighter. In 1954 it was the only wholly new British military jet at Farnborough and it was entered for that year's NATO light-fighter competition. The competition was based on Petter's concept but included the requirement to use low-pressure tyres in order to allow the fighters to operate off short, unprepared strips. The Gnat, however, was designed for hard runways and Petter refused to modify the undercarriage – resulting in the Folland being eliminated from the competition in 1955, even though it was the only competitor to meet the competition's required weight.

In spite of internal opposition, the Air Ministry approved production of six Gnat F Mk1s. Following the first flight on 16 May 1956, development continued with little trouble, in contrast to the problems besetting the Hawker Hunter. When the RAF indicated they would consider a trainer version, Follands provided the relevant drawings and documentation. Approval was given for the development of the Gnat TMk1, which was fitted with a larger engine, a different wing and tail, revised control surfaces and a twin-seat tandem cockpit under a large clear canopy. The design changes created some inflight instability, which was overcome through a complex set of gears – challenging to many a pupil pilot.

Only after Folland had been absorbed by Hawker-Siddely in 1959 did the first T1 fly. Teething troubles were addressed by incorporating the improvements being made to the Indian and Finnish F1 models and serviceability improved.

All operators found – ironically – that the small size and therefore expected ease and low cost of maintenance was not experienced in practice. The confined working space and concealment of many of the fittings simply made servicing difficult and injurious. Taller pilots were known to have been injured when ejecting, while instructors found the rear cockpit cramped and the forward view restrictive.

The Gnat served for sixteen years as a primary trainer but, due to its special features, there was a great deal a pupil had to learn simply to fly the Gnat that they would never use in any other aircraft. They were replaced in 1978 by the Hawker-Siddeley Hawk, but continued flying in limited roles until 1980.

Foreign service

Early on in the development phase, the IAF showed keen interest and ordered forty aircraft, fifteen of which were to be supplied in kit form. Hindustan Aircraft Ltd (HAL) then began licenced production of more than 200 Gnat units and developed the airframe further, creating the HAL Ajeet ground-attack aircraft. A naval version for the Indian Navy proved unsuitable as the Gnat was too light for the steam catapult aboard INS *Vikrant*.

IAF Gnats performed superbly as fighter aircraft in the two Indo-Pakistani wars. In the 1965 war they were labelled 'Sabre Slayers', being credited with downing seven enemy F86 Sabres, for the loss of two shot down and one captured. Success was attributed to its agility and small size, which made it difficult to see in the heat haze of the combat zone. Again in the 1971 conflict they faced the Sabres. Most famously, on 22 November 1971, four Gnats were scrambled to drive off four Sabres, shooting down three. The Gnats served the IAF well until 1978.

For the Finns, the Gnat was a key component of their air force's capability as it was their only high-performance jet fighter and they purchased thirteen aircraft. These remained in service until 1974 and one was the first aircraft in which a Finn 'broke' the sound barrier. It received a mixed reception and, after two crashes, it was viewed as unreliable but, on the whole, the pilots liked it and overcame the instabilities; the two reconnaissance airframes proved very useful.

Only two trial examples were accepted at the Yugoslav Air Force Air Test Centre in mid-1958, where one was written off in a landing accident. The Yugoslavs found the Gnats difficult to maintain and, when the costs of building them under licence proved too expensive, no further orders were forthcoming.

The Gnat became a household name through its service with two famous aerobatic teams. The first was the Yellowjacks, formed in 1964 – the forerunners of the Red Arrows. They were based at RAF Kemble for thirteen years and the nimble little jet helped to create the expertise and mystique that the Red Arrows enjoy today.

Hunting Jet Provost

Arguably the most successful early jet trainer was the Hunting Jet Provost. Hunting had diversified from shipping into aircraft maintenance and transport in the 1930s, acquiring the famous Percival Aircraft Company in September 1944.

Early jet pilots would have some pilot's notes and cockpit instruction from an experienced pilot and then they would simply set out on their first flight. Some aircraft, such as the Meteor, were developed into trainer versions but no dedicated introductory jet trainer aircraft existed. The RAF's requirement for such an aircraft saw Hunting take the piston-powered Percival Provost and develop it into the Jet Provost. The prototype Hunting Percival P 84 Jet Provost XD674's initial flight was from Luton on 16 June 1954. It had the standard Provost wings attached to a fuselage, which retained the Provost side-by-side cockpit in a fuselage; this, in the rear section, housed an Armstrong Siddeley Viper ASV.5 (1,640 lb thrust) engine. This aircraft can be seen at RAF Museum Cosford.

An order for ten service-test Jet Provost T1 units was placed in 1953. Conceived as a basic jet trainer, it retained the handling qualities of its piston-engined forebear. Tests with the TI showed that flight instruction was reduced from 240 hours training on piston aircraft followed by the Vampire to 167 hours purely on the Jet Provost. The trainer was designed to produce docile flying characteristics and good training qualities, but to represent the handling of operational aircraft.

During the evaluation stage, more than 2,000 hours were logged and a number of cockpit layout and other amendments were suggested, which resulted in the T2 version. This had the more powerful Viper ASV.8, which allowed for a more aerodynamic rear fuselage and required a slightly enlarged jet pipe. The long-legged undercarriage of the Mk1 was replaced by a shortened gear; hydraulic control systems were provided and the cockpit canopy improved to take advantage of the field of vision allowed by the short nose.

Four Mk2s were built for development and demonstration flying, and were also used for promotional tours in Europe, Canada, the United States and Latin America. *Flight* magazine reported that these aircraft were 'flying up to eleven sorties a day in the hands of numerous pilots, and only the barest minimum of attention being necessary to either engine or airframe'.

In June 1957 the RAF placed an order for forty Jet Provost T3 aircraft to be the standard *ab initio* basic trainer and Hunting went into full production at their Luton factory. This was the first jet trainer purchased by any air force as a standard primary trainer. The pupil and instructor sat side by side under a sliding clear-view canopy in Martin-Baker Mk4 ejection seats, with the radio and electrical equipment easily accessible in a compartment under a hinged nose cone. Humphrey Wynn wrote in *Flight* magazine in April 1960 that,

> once upon a time, all the pupil needed to do when he got into a training aircraft was to clip on his Sutton harness and plug in his speaking-tube. Nowadays life is much more complicated, and I am generally amazed at the amount of plugging-in and fastening that has to be done.

Between 1958 and 1962, 201 T3s were delivered as the RAF flying schools progressively exchanged their piston aircraft for jet trainers. The T3 and T3A avionic upgraded version served the RAF in the training role for thirty years.

Development of the Provost continued in the form of the T4, which made its first flight in summer 1960. Changes included the more powerful Viper 201 engine, and minor airframe and equipment upgrades. Delivery to the flying schools began a year later, while other RAF units used the T4 for specialised training.

Increased engine power meant that the aircraft were subjected to far greater stresses than the T3 and they accordingly suffered the early onset of airframe fatigue. Major budget cuts and the widespread closure of flying schools meant there was little need to consider a refurbishment programme and the T4s were retired far earlier than planned.

Continued development in jet flight meant that the RAF increased its demand for high-altitude training and unpressurised T3s and T4s could not meet this requirement. The British Aircraft Corporation (BAC), which had taken over Hunting-Percival in 1964, developed the T5 to meet this requirement. It was fitted with a more powerful engine and the basic T4 cockpit and forward fuselage were redesigned to accommodate pressurisation. After evaluation, the RAF ordered 110 units, with the first delivery in 1969.

Display teams

Aerobatics and flying schools went together, and never more so than when equipped with the Jet Provost. This was the first jet in the new world of aviation that was suited to formation aerobatics and stable enough to be flown in intricate manoeuvres. More RAF aerobatic teams flew the Jet Provost than any before or since, and every flying school had a number of different display teams. The founders of one of the most famous were the Sparrows of 1958, followed by the Redskins from the Central Flying School at RAF Little Rissington. From these teams emerged the Pelicans flying the T3, initially in standard colours; however, when re-equipped with the T4 in 1962, they adopted a bright-red colour scheme and were renamed the Red Pelicans. In 1963 they extended their formation from four to six aircraft and became the premier RAF display team, performing around the United Kingdom and in Belgium and France. They in turn were replaced for the 1965 air show season by the Red Arrows flying the sporty Folland Gnat. The now four-ship Red Pelicans continued to perform until 1973.

In foreign service

This robust, versatile trainer proved popular in foreign air forces. The T3 version was exported as the T51 and supplied to Ceylon, Sudan and Kuwait.

Twelve were sold to the government of Ceylon, fitted with two .303 machine-guns, four rocket rails and racks for eight 25-lb bombs, making it ideal for a small air force requiring light ground-attack aircraft. The tandem seating allowed the pilots to share the workload and a reflector gunsight was fitted in front of each seat. The aircraft had exceptional handling qualities and the two-hour flight endurance at sea level with a full rocket load provided operational flexibility. Taken into service by the then Royal Ceylon Air Force in 1959, they were phased out by the Sri Lanka Air Force only in 1979.

A natural development of the more powerful T4 was the T52, of which Iraq purchased twenty, Venezuela fifteen and the Sudan added eight to their T51 fleet. Sales were disappointing and only five T55s – the export version of the T5 – were sold to Oman.

However, the concept of a light-attack aircraft had been proved. Development of the T5 and the subsequent very successful Strikemaster falls into the category of third-generation jets.

The Folland Gnat T.1 prototype, XM691, at Farnborough in 1960. It was always used for tests and development until withdrawn in 1964. Purchased by Donald Campbell, the tail, engine and other parts were used in building his record-breaking Bluebird K7 boat. This boat crashed in January 1967 but was recovered in 2001 and is under restoration with the Bluebird Project. (John Kendall)

1963 Folland Gnat T.1
XR540, built in 1963,
seen here at RNAS
Culdrose in 1983 in Red
Arrow colours. (Andy
Chetwyn)

Line up of ten Red
Arrow Gnats at an air
show in the late 1970s.
(George Woods)

Gnat XR572 was built
in 1963 and joined
the Red Arrows as the
spare in June 1974. Its
final flight was in 1979,
now privately owned
in the United States.
(George Woods)

Classic Red Arrows
diamond formation.
(George Woods)

Gnat T1 XS102 painted
to represent the lead
aircraft XR992 of the
'Yellowjacks' team, on
display at Fairford in
2014. (Jonathan Francis)

A great view of the
underside of XS102
during a display at
Kemble Air Day in June
2008 at Kemble Airport.
(Adrian Pingstone)

Gnat T1 XP504, in 1960–70s RAF 'Red Arrows' display team colours and marked as XS111, performs at Cosford in June 2012. (Jonathan Francis)

Gnat T1s XR538 (G-RORI), XS102 (G-MOUR) marked as XR992, and XP504 (G-TIMM) marked as XS111 at RAF Waddington, 2014. (Jonathan Francis)

Indian Air Force Gnat E1070 on take off. (Bharat-Rakshak.com)

A group of Indian Air Force Gnat pilots. (Bharat-Rakshak. com)

Replacement of the tall undercarriage, improved access to the tail pipe and a more powerful Viper 8 engine were features of the T2. Here G-AOUS, one of only four T2s, taxis past a row of T3s at Luton. (Jeff Morgan)

Delivered in 1961, this T3A XM479 was one of the last T3s to be upgraded to T3A standards in 1976. Sold to a private owner as G-BVEZ in 1993, it appears on the show circuit, as seen here at Kemble in 2009. (Kevin Slade)

6 FTS Acklington T3 XN503, seen here in 1965, survives in part. Serving various flight training schools from 1960 until 1976, the cockpit section was used for ground instruction and is now with the Boscombe Down collection. (Niel Aird)

This T4 XR679 is camouflaged, as used for training by No. 1 TWU/79 Squadron Brawdy. Four of these T4s were used to give Army Forward Air Controllers air experience, while the aircraft were used by the controllers for initial ground-attack exercises. (Rob Schleiffert)

XM424 XM466 Hunting Percival Jet Provost T3A at RAF Boscombe Down in 1992. (Kerry Taylor)

On show at RAF Waddington in 2014 is the now privately owned Provost T4 G-BXLO ex-XR673 in the No. 2 FTS colour scheme in which it was delivered in 1963. (Jonathan Francis)

The Red Pelicans of CFS Little Rissington were in 1973 flying Provost T5 aircraft: XW291, XW288, XW293 and XW289. (BAE SYSTEMS)

It was only for the 1962–64 seasons that the Provost-equipped Red Pelicans existed. During the end of the 1964 show season, they gave co-ordinated displays with the RAF's first Gnat-equipped team, the 'Yellowjacks' from No. 4 FTS at Valley. (Robert Haigh)

Fighters

Early second-generation jets were designed when less was known about swept wings. Today's designs have made these aircraft considerably safer through advanced understanding of aerodynamics, computer management to improve handling and engines with swifter reactions and far more power to get a pilot out of trouble.

Supermarine Scimitar

The last aircraft to bear the Supermarine name, the Scimitar, replaced the Sea Hawk primarily as a ground-attack and strike aircraft. Its armament comprised four 30 mm Aden guns and it was also capable of carrying a tactical nuclear bomb. Evolved from a series of twin-engined development aircraft that Supermarine had used to explore jet power and aerodynamics, the Type 525 (first flown in April 1954) became the forerunner of the Scimitar.

The Royal Navy placed an order for 100 aircraft, powered by twin Rolls-Royce Avons. Noteworthy was the flap-blowing system, where air is bled off from each engine and blown aft from a supersonic nozzle over the surface of the flap, then downwards and aft. The deflection of a large mass of air almost doubles the effectiveness of the flap and reduces the landing speed, resulting in a more nose-down attitude on approach and improved stability; all necessary for landing on a carrier. The air supply from each engine was also interconnected to overcome asymmetric flying conditions in the event of engine failure.

A special testing unit, 700X Flight, was created in order to evaluate and develop ways of operating the new twin jet ahead of its introduction into service. The first Scimitar unit was No. 803 Squadron FAA, formed in 1958, which embarked eight aircraft onto HMS *Victorious* in September for tests – it proved reliable, robust and was liked by the pilots.

As with all the early jets, there were no dual trainers available and computer simulators were a long way off. John Middleton's story was told in the July 2007 edition of *Aeroplane*:

> Your first solo in the aircraft was also your first flight. There was no two-seat trainer variant, but there was a dummy cockpit in which to practise procedures and learn the layout (it is now at the City of Norwich Aviation Museum — www.cnam.co.uk). After that you walked out to the aircraft. An instructor stood on the cockpit entry ladder and watched John do all the preliminaries. He then patted him on the head and said, 'Enjoy yourself'.

Middleton noted that everything happened 'ten times faster than on a Sea Hawk'. It was a high-speed ride, taking 80 seconds to reach 5,000 feet and capable of 710 mph or Mach 0.968 at sea level.

This extreme power was almost the undoing of Lieutenant K. Brent Owen of No. 803 Squadron, as he revealed in details of his first flight in a Scimitar:

> There being no dual-control Scimitars, one's first flight was a lonely affair. The big problem seemed to be the fierce acceleration, so I would tame the beast by the simple expedient of using only partial power. Anyone who knows anything about flying will not be surprised at the result. I found the resulting violent wing-drop at lift off highly stimulating ... and after firewalling both throttles it was indeed 10,000 feet before my eyeballs stopped spinning and I got the gear and flaps up.

Its huge power, astounding acceleration and innovative aerodynamics did not save it from being a failure. The advanced ideas incorporated in its design contributed to a lack of success in service. Of the seventy-six built, 50 per cent were lost in accidents. It leaked fuel and oil, suffered inflight fires and, although below 20,000 feet it could out-fly anything, at altitude it was unstable. Like the Supermarine Swift, a slight turn at height resulted in the nose pitching up followed by a spin. This was caused by the compressibility problem, where airflow characteristics change as the speed of sound is approached, which leads to a high-speed stall at the wing tip and an increased flow downwards over the tailplane. These issues were overcome to a certain extent during development, with the introduction of a sawtooth leading edge to the wing, boundary layer fences and Küchemann wingtips. Here a long curve ran from the sawtooth to the leading edge to tip but, unusually for this type of wing tip, on the Scimitar the trailing edge was square cut. In spite of these efforts, many pilots were lost when, on final approach, the high drag effect caused the nose to pitch up and the aircraft stalled too low for recovery.

The handling difficulties were exacerbated by the power and weight of the Scimitar, which was out of proportion to the size of the current Royal Navy aircraft carriers. While the Americans were building larger ships, the British Navy was under increasing financial pressure and on the smaller carriers there was absolutely no room for any error.

In addition to its Aden guns, the Scimitar could fire Bullpup air-to-ground missiles, Sidewinder air-to-air missiles, unguided rockets and drop four 1,000-lb bombs. It could increase its range with four 200-gallon drop tanks or carry an inflight refuelling pack, used for buddy refuelling. Later, while in second-line service, they refuelled their replacements, the Buccaneers, which could not take off when fully armed and fuelled.

Never firing a shot in anger, the Scimitar served only in a classic British 'Gun Boat' diplomacy manner: first, in 1961 in the Gulf they were integral to the naval force positioned to prevent the Iraqi invasion of Kuwait; and then No. 803 Squadron served two tours on the Beira patrol to enforce oil sanctions against the rebel government in Rhodesia.

However, they produced some exciting performances at the annual Farnborough Air Show. The Scimitars of No. 807 Squadron FAA excelled themselves at the 1959 show. Led by Lieutenant Commander Keith Leppard RN, they flew in formation with a Vulcan; one picked up a target with its modified arrestor hook and the aircraft performed the 'twinkle roll' for the first time. Each one performed a rapid individual roll while flying in an open box formation, described by *Flight* magazine as what 'appeared as some fantastic piece of gambolling by a school of dolphins'. The final 'follow-my-leader' low pass by the five Scimitars produced four audible near-sonic almost-bangs.

Development of the more capable two-man Blackburn Buccaneer heralded the end of frontline service for the Scimitar from 1966. They were used for inflight refuelling, target-towing, radar trails and training. The last flight of a Scimitar was in December 1970, when the civilian Fleet Requirements Unit ceased using the fighter.

Hawker Hunter

Work started on one of the greatest British fighter jets in 1947 when Hawkers developed the P1067 fighter with all swept surfaces and powered by a Rolls-Royce Avon engine. In June 1948 three prototypes were ordered, two powered by the Avon and one by an Armstrong Siddeley Sapphire engine. A number of design changes were made, including the introduction of a four 30 mm Aden gun armaments pack. This complete unit provided exceptional combat capability, as the pack could be demounted and rapidly replaced.

An order was placed for 113 FMk1s and later for the Sapphire-powered FMk2. The Hunter flew for the first time in the hands of famous wartime and test pilot Neville Duke on 20 July 1951. Extensive trails and adjustments were made, leading to the first production Hunter FMK1 being flown in May 1953.

After a year of testing various fittings, wing design and aerodynamic ideas, No. 43 Squadron RAF was the first operational unit to receive the Hunter. In spite of its lasting success, the Hunter experienced a number of problems in early service. Production had been rushed through as part of the Priority System, which saw important military developments fast tracked; a consequence of this was that the early versions of the Hunter were not really fit for operational service. The Avon engines tended to surge and cut out when the guns were fired at altitude and their use at height was restricted. As with most early jets, fuel consumption was poor and, accordingly, effective operational range limited.

Although the Sapphire-powered FMk2 did not suffer flameouts when the guns were fired, only a small number were built before the similarly powered FMk5 was delivered from 1955. This variant equipped five squadrons and took part in the Suez Crisis, but developments in the Avon engine saw it selected as the engine of choice and the FMk5s only served until 1958.

A single FMk3 WB188 was built with an Avon RA7R with reheat. Painted bright red and flown by Neville Duke, it set an Absolute Speed Record of 727.6 mph in September 1953.

Tests on the FMk1 had been flown with drop tanks and it was decided to increase internal fuel capacity and provide a pair of underwing pylons to carry drop tanks or armaments. This became the FMk4, which, after the first delivery in 1955, went on to equip twenty RAF squadrons. These were the first day fighters to be equipped with radar and powered controls.

In a further development, a more powerful Avon 200 engine and another pair of pylons were added to create the FMk6. This suffered from the pitch-up problem created by the 'compressibility problem' typical of the early jets as they approached the 'speed of sound'.

A dog-tooth wing plan, which had extended outer leading edges, overcame this limitation and the FMk6 became the primary RAF interceptor.

By 1960 the Hunter's effectiveness was being eroded by the speed of technological developments. More advanced fighters were in production, the large bombers could reach altitudes unobtainable in an F6, and most importantly, the manned gun-equipped fighter was set to be replaced by an integrated weapons system, with a missile replacing guns.

Typically for the era, the first trainer was developed only after the frontline squadrons had received their Hunters. The TMk7, based on the FMk4, was a twin-seat trainer with the instructor and pupil seated side by side. First flown in July 1955, it required almost a year of modifications to overcome the instability caused by the new cockpit arrangement. Most Hunter Squadrons had T7s on strength for pilot rating and conversions, these serving into the 1970s until replaced by the Hawk.

A development of the T7 was the T8 built for the Royal Navy. The Vampires were no longer suitable for training pilots to fly high-speed aircraft such as the Scimitar. The T8 undercarriage was not modified to take the stresses of carrier landings, but it was equipped with an airfield arrester hook used for land-based arrester gear practice.

Serving with RAF Strike Command from 1960 was the FGA (Fighter Ground Attack) Mk9. These were all converted Mk6 aircraft that were originally designed for use in the Middle East and foreign bases. The cockpits were air conditioned, wings were strengthened, a braking parachute added, additional strong points for armaments installed and an increased fuel load carried in the new 230-gallon drop tanks stressed for combat use. Peter McLeland reflects:

> Well the Venoms have reached the end of their 'Fatigue Life' and they have been scrapped. We all flew home to England in a Britannia and collected a full set of new Hunter FGA9s. We flew them back out from England, routing Stradishall, Luqa (Malta), Nicosia (Cyprus), Teheran (Persia), Sharjah (Emirates), Khormaksar (Aden), Embakasi (Kenya) ... Eastleigh was too short for Hunter operations. The flight out from England to Kenya took us three days.

After about nine years the Mk9 was replaced by the Phantoms, Jaguars and the Harrier, although they were used as weapons trainers into the mid-1970s.

A replacement was sought for the Supermarine Swift FR5 to continue the reconnaissance of the 'Communist Bloc' from Germany. Three forward-facing cameras were fitted to thirty-five FMk6 airframes – known as the FRMk10 – and, like many Hunters, it served until the end of the 1960s.

The GAMk11 were 40 FMk4s where the gun pack was removed and additional underwing storage options were provided. These were used by the Fleet Air Arm as weapons trainers.

In combat

Of the second-generation fighters, the Hunter saw the most action across British areas of influence. After rather periphery engagement in the Suez Crisis, it was used to escort transports during the 1958 Iraq revolution and immediately thereafter to prevent an Iraqi invasion of Kuwait.

None of these roles had required any shots to be fired in anger. This all changed with the low-key war in the Radfan area of Aden. A series of attacks on British forces were countered when FGA9s carried out rocket and bomb attacks on the insurgents and provided air support to SAS forces. Intensive action on the ground and in the air defeated the rebels.

The FGAMk9 acted in a similar role during the Malaysian Confrontation, patrolling against the Indonesian air force and carrying out rocket and gun attacks against rebels and Indonesian troops.

Foreign service

Like the Canberra, the Hunter was an exceptionally successful export product, used by nineteen countries and seeing extensive combat in many regions and used until the early 2000s.

In Europe, the Dutch and Belgians produced the Hunter under licence, starting with the FMk4, then the FMk6 and the Dutch-built TMk7, which they shared with the Belgians. On retirement in the 1960s, many of these were purchased by Hawkers and refurbished and sold on to other smaller air forces. Sweden operated the F50 and the Danes the F51, both broadly similar to the FMk4 and the T53, the export version of the TMk7.

The most famous European operator was the Schweizer Luftwaffe (Swiss Air Force), which ordered 100 F58 fighters and then, in the 1970s, a further fifty-two along with eight F68 trainers. In the 1970s another sixty Hunters were purchased. The fighters were converted FMk6s upgraded to FGA9, but armed with Sidewinder missiles, a larger fuel capacity and anti-missile defences. Along with the popular Patrouille de Suisse aerobatic team, the Hunters were retired only in 1995, with many still flying in private hands.

Engaged in a significant expansion programme in 1957, the IAF ordered 160 FMk56s: a large number were Dutch and Belgian built and twenty-two were new-build T66 trainers. These were supplemented by fifty-three FGA56As and seventeen T66s all refurbished, delivered in the mid- to late 1960s. Like the Canberra, they saw their first service over Goa and then first combat in the first Indo-Pakistani war of 1965, where they were used mainly for ground attack and troop support. During the second war in 1971, the Hunters were again used to support troops, experiencing some losses. Following their replacement by more modern equipment, they continued in dwindling numbers in various roles until almost the end of the twentieth century.

A fleet of overhauled Hunters were operated by Singapore for twenty years. In 1970, twelve Hunter FGA74s, similar to the FGA9s, and four FR74 reconnaissance aircraft were delivered and then added to with another twenty-two FR74Bs and eight T75 trainers. Some local modernisation took place, which included fitting with missiles.

Hunters were versatile and, like most air forces, Chile purchased, from 1968, a mixed fleet of reworked Hunters. The majority were FGA71, supported by T72 trainers and FR71A reconnaissance aircraft. When the military regime came to power in 1974, they were used to attack the opposition, but became largely unserviceable after Britain cut off spares supplies due to the government's poor human rights record. Post-1982, support for Britain's Falklands campaign led to twelve FGA9s and two ex-Kuwaiti T67s being supplied. Subsequently, operational aircraft were upgraded with a local radar, anti-missile defences and the ability to carry and fire air-to-air missiles.

F52, standard FMk4s were sold to Peru in 1957 and a T62 trainer three years later. With the purchase of the Dassault Mirage 5, the Hunters were obsolete as primary fighters and they were assigned a ground attack role – only being retired in 1980.

Although purchased in smaller numbers, the Hunter was popular in the Middle East and Africa, where it served in Abu Dhabi, Kuwait, Somalia and Iraq, where they were used in the 1967 Six Day War and the 1973 Yom Kippur War against the Israeli Air Force. The twenty-two Jordanian Hunters claimed to have shot down Israeli aircraft in earlier conflicts but they were all destroyed on the ground in the Six Day War. These were replaced from 1968 with reworked FMk6s and, by 1972, a total of thirty-eight Hunters were in service. Abu Dhabi donated their surviving aircraft to Jordan, who in turn donated all the surviving thirty-two aircraft to the Sultan of Oman in 1975, where they were finally retired in 1973.

Lebanon obtained a total of nineteen Hunters from various sources from 1958 into the mid-1970s, and they were used in various skirmishes until the late 1990s. Four were returned to service in 2008, but appear to have been replaced by Hawks obtained from the UAE. Qatar and Saudi Arabia operated a limited number of Hunters in interim roles until replaced by more modern aircraft.

The RAF and the Royal Rhodesian Air Force (RRAF) were closely linked through their ethos and the fact that many RRAF personnel had served in or were trained by the RAF. A visit to Rhodesia by No. 8 Squadron FGA9s in January 1960 so impressed the Rhodesians that, three months later, they ordered twelve of the fighters. Delivered between 1960 and 1963, they became the frontline aircraft during the 1970s civil war. RhAF FGAMk9s were modified and maintained locally, delivering a variety of weapons and also providing a reconnaissance role with the adaption of a 100-gallon fuel tank as a camera pod, fitted with a single forward-facing and two oblique-facing units. Only three were lost; two to ground fire but, after the war, a further five were blown up at the Thornhill air base. These were replaced in the 1980s, finally being retired in 2002.

Further north in Kenya a new air base was built and the air force purchased its first jet combat aircraft, six FGAMk80s and two TMk81s. In 1981 the seven survivors were handed to the Air Force of Zimbabwe.

English Electric Lightening

The volatile William 'Teddy' Petter had begun work on a supersonic aircraft in 1948, but his involvement with the Canberra saw him hand the design lead to Freddie Page, who, with

Petter's sudden departure from English Electric, then became Chief Designer. The aircraft he developed was known as the P1. Three prototype research aircraft were built, capable of reaching Mach 1.2 but with the long-term plan of manufacturing operational aircraft. The P1 was designed with a radical 60-degree swept wing and powered by two 7,500-lb Armstrong Siddeley Sapphire SA5 engines mounted one above the other in order to remove the danger of asymmetrical flight if one failed. The first P1 WG760, achieved level flight supersonic speeds in August 1954, piloted by Roland Beamont:

> The P1 was delightful to fly. I had a very good first flight in it. It went on so fast that we had it transonic on its second flight and supersonic on its third flight, in level flight. The first time over Britain, in a British aeroplane in perfect fingertip control, it was delightful to fly and I went barrelling down the Solent on an August day at about 40,000 ft with the Mach meter at about 1.02 or 1.03.

During 1958 the first of the proposed supersonic fighters, the P1b, powered by Rolls-Royce Avon 200R engines and with double the thrust of the Sapphires, was the first British aircraft to reach Mach 2, a significant achievement a mere seventeen years after the first UK jet-powered flight.

In late 1956 an order for twenty of these aircraft was placed so that testing of every aspect of the new fighter could be accelerated. The result was the Lightning F1 Mach 2 Interceptor, which entered service with the RAF in 1960. It was lucky to escape the cuts outlined in the White Paper by Duncan Sandys. This document postulated that the age of the manned combat aircraft was over and that future defence and attack capability would be provided through missile technology. Many ground-breaking designs were scrapped, while the Lightning was accepted as an interim measure until the necessary missiles were delivered. As a consequence, any development was rejected and the F1 remained limited in range and capability, which allowed the American manufacturers to fill the needs of European air forces. As with all new aircraft, there were serviceability issues; very few hours were flown and they were grounded for three months, but these issues were overshadowed by the sheer power of Britain's first supersonic fighter.

With the surface-to-air missile (SAM) technology behind in development, an order was placed for an F2 variant with a more powerful Avon 210, with variable reheat functionality and better instrumentation and radar.

By this time the TMk4 side-by-side trainers had been delivered, which made conversion to the super-fast Lightning a lot easier.

Further minimal modifications were found in the F3, which began to replace the F1 and F1A. This variant was fitted with a squared-off vertical stabiliser, ventral fuel tanks to extend its rather poor range, a more powerful engine and improved wing leading edge for enhanced performance. In addition, it was fitted with improved radar and, in a ground-breaking step, the cannons were removed and it became a pure missile-armed aircraft. Somewhat unusually, the Lightning was provided with above-wing string points to carry auxiliary fuel drop tanks.

Developed alongside the F3 was the T5 twin-seater trainer, which incorporated all the improvements of the fighter version. One significant change was that the central throttle quadrant was located to the starboard side, which was opposite to the fighter, and pilots had difficulty in adjusting when flying from the right seat.

A long-range version the F6 had an area ruled under fuselage tank and additional fuel storage in the flaps, which increased the range by 40 per cent. It was also fitted with a refuelling probe and the ability to carry the overwing tanks. Armed with cannon and able to fire the Red Top missile, the wing was larger with kinked edges. Some F2s incorporated these changes and became the F2A model.

First discovered in Germany during the war and then proved in a wind tunnel by Richard Whitcomb, the Area Rule became a critical design factor in jet aircraft. As an aircraft approaches the 'sound barrier', the shock waves formed at various points across the airframe can greatly reduce power due to what is known as 'wave drag'. Whitcomb discovered that, if he pinched the fuselage at the wings and other points where the air bunched and then smoothed out its flow across the cross-section of the aircraft, this reduced drag and increased speed. Although today's military and commercial jets still adhere to some of these principles, modern jet engines can effectively power through the pressure of the air build up, allowing for greater design freedom.

From 1966 pairs of Lightnings were based at QRA stations, ready to scramble and escort Soviet aircraft out of the area as they probed the UK defences. Pilots would sit in the cockpits for two hours at a time, 24 hours a day and could be in the air at 30,000 feet within about 3 minutes. The renowned fuel-limited flying time of the Lightning was alleviated by the F6 modifications, but also by the inflight tanking provided by the converted Victor bombers.

When the replacement Tornado multi-role fighter was behind schedule, the RAF were forced to extend the service life of the Lightning until April 1988 when the Tornado Air Defence variant came on strength.

The only export customers were Saudi Arabia and Kuwait. Between them they purchased forty F52 and F53 fighters and T55 trainers. The Kuwait versions were suffixed with the letter K but differed little from those supplied to the Saudis. The F53 was really an F6, but with ground-attack capability through rocket pods and bombs being fitted on both upper and lower wing string points. Roland 'Bee' Beamont was one of the pilots involved:

> One of my jobs was to take part in the initial stages of exporting the Lightning to Saudi Arabia, who hadn't got a supersonic air force and were negotiating with the Americans to take them into the supersonic era. We thought nothing ventured, nothing gained, so we went and challenged the Americans who were offering the F-104 Starfighter with the Lightning. ... with the help of a fine demonstration by my colleague Jimmy Dell who took an air force Lightning in there, we got the contract to supply a force of Lightnings and a force of Strikemaster jet trainers and all the basics of flying training.

Only the Saudi F52s, refurbished F2 Lightnings, and two F54s saw any action. Rushed out to Saudi in 1966 to counter Yemeni border violations supported by the Egyptian Air Force, the Lightnings ground-attack operations ended the conflict quickly and effectively.

The cockpit of Scimitar F Mk1 XD232 a month after its first flight, in 1958. It was successively assigned to three navy squadrons, including the Fleet Requirements Unit and then the Foulness Island Ranges. The remains were scrapped only in 1984. (BAE SYSTEMS)

Two of these Rolls-Royce Avon 202 turbojets powered the Scimitar. The design team behind Rolls-Royce's first axial flow engine was headed by A. Cyril Lovesey, one-time head of design of the famous Merlin. From 1950 the Avons powered aircraft such as the Canberras, Comet, Swift, Hunter, Sea Vixen and Lightning. Throughout its twenty-four-year production history it was uprated to meet the demands of the fast-changing world of the jet. (BAE SYSTEMS)

Aboard HMS *Eagle* in 1963: maintenance being carried out on Scimitar XD280 of No. 800 NAS. Delivered in 1959, this aircraft served until 1971, ending up as a ground trainer. (Andrew Patterson)

Two unidentified Scimitars preparing to take off from HMS *Eagle* while the Westland Whirlwind rescue helicopter stands by. (Andrew Patterson)

Crew rush to unhook the No. 800 NAS Scimitar F1 and the ever-present Westland Whirlwind rescue helicopter is on duty to rescue any pilot who misses the arrestor wires. (Andrew Patterson)

While operating off Malaysia in 1965, Scimitar F1 XD328 R015 of HMS *Ark Royal's* No. 803 Squadron, FAA, was written off. The barrier-rigging team erected the barrier in the record time of 2 minutes, 34 seconds. (Peter McLeland)

Deck crew work to make XD328 safe after the barrier landing. The aircraft was sent to RNAY Fleetlands in November 1965 for repairs but they proved to be uneconomical, so it was scrapped a few months later. (Peter McLeland)

Scimitar F Mk1 XD215 is rested on its tail bumper for take-off, the nose wheel high in the air. This increased angle of attack was necessary to compensate for underpowered catapults and to bring the heavy Scimitar to flying speed. (BAE SYSTEMS)

The beginning of the take-off sequence, the steam catapult having started to run. Refitted with an underwing refuelling pod, aircraft will attain a speed of about 150 mph by the end of the 165-foot track. (BAE SYSTEMS)

Hawker's famous chief test pilot, Neville Duke, was integrally involved in developing the Hawker Hunter. (BAE SYSTEMS)

First flown in 1952 and the first Hunter to have the Aden gun fitted, WB195 was the second prototype. (Jet Age Museum)

Indian Air Force Hunter F56s in the United Kingdom, prior to delivery. (Bharat-Rakshak.com)

Built as an F6, this Hunter was first modified to Mk9 standard in 1959 and later converted to FGA9 standard, served in the Aden Strike Wing until the final withdrawal of British forces from the region. In 1984, it participated in the final formation of single-seater Hunters and XG154's final flight. Eventually assigned to the RAF Museum at Hendon. (BAE SYSTEMS)

Hawker Hunter F Mk6 XF516/G-BVVC, operated by Hunter Flying Club, taking off from Chivenor in 2000. Written off in 2003 after electrical failure led to an engine flame-out. (Jonathan Francis)

A mixed formation of No. 79 Squadron FGA9s with T7s all part of No. 229 Operational Conversion Unit at RAF Chivenor, which flew these between 1967 and 1974. (BAE SYSTEMS)

Seen at Fairford in 2002 masquerading as the Hunter prototype WB188, this F6 was WV256 delivered to the RAF in 1955, transferred first to the Royal Navy after conversion to a GA11 in 1963, and then to Fleet Requirements and Air Direction Training Unit (FRADTU) in 1972, remaining in service until 1995. Seen here repainted in its original RAF No. 26 Squadron marking as WV256. (Jonathan Francis)

Hunter T7 G-BXKF (ex-XL577) seen here on take-off from Kemble in 2008 with T7A G-FFOX (ex-WV318, built as a F4 but converted to a T7 in 1959). (Kevin Slade)

Hunter WV267 was built as an F4 for the RAF and then transferred to the Royal Navy. Converted to GA11 standard in 1963, it flew first with the RN Airwork's Fleet Requirements Unit (FRU) and then (as seen here) with Fleet Requirements and Air Direction Unit in 1985, four months before it was grounded following a bird strike. (Rob Schleiffert)

XL580 was the first production T8 Hunter built for the Royal Navy. Seen here in around 1964, operating from RNAS Yeovilton as the 'Admirals Barge' – the flag officer of Flying Training's personal aeroplane. It was modified with a Harrier-type cockpit, nose and radar for familiarisation training. Now on display at the FAA museum. (BAE SYSTEMS)

A Hunter F60 prior to delivery to the Royal Saudi Air Force (RSAF) in 1966. The RSAF gifted it to the Royal Jordanian Air Force in 1968. (BAE SYSTEMS)

Hunter J-4058 is a 1959
Mk58 of the Swiss Air Force,
seen at RAF Waddington
in June 2004. Purchased by
Hawker Hunter Aviation.
With its military registration
ZZ191, it is flown on defence
simulation tasks. (Jonathan
Francis)

Delivered to the Swedish
Air Force in 1957, converted
by Hawker Siddeley to a
TMk68 in 1970 and sold on
to the Swiss Air Force as
J-4206. Now privately owned
and run as HB-RVV by the
Hunter Flying Group. (Kevin
Slade)

Chilean Hunter FGA Mk71s,
part of the first order of the
aircraft refurbished to FGA
Mk9 standard. J705 is an
ex-Dutch FMk6, delivered
in 1967. Next in line is
ex-RAF XG232, delivered
in 1968. (Museo Nacional
Aeronautico y Espacio Chile)

A Rhodesian Air Force RhAF FGA 9 landing with its braking parachute. (John Frame www. Retributionbook.com)

But for the background and the single assagai overlaying the RAF-style roundels of this FGA9 of the Rhodesian AF, this could be an RAF Hunter. (John Frame www. Retributionbook.com)

F6A ex-RAF XF515 G-KAXF at Kemble in 2011 in Dutch markings as 'N-294', on the European show circuit. (Kevin Slade)

The now-defunct Team Viper Hunter display team led by T7 'WV372' (G-BXFI), with FGA9 'XE601' (G-ETPS) top, GA11 'XE685' (G-GAII) below, and T7 'XL600' (G-VETA) top with PR11 'WT723' (G-PRII) below in Black Arrows markings as 'XG194'. (Kevin Slade)

The second prototype WG760 was designated a P1A, and was essentially the same as the first prototype but fitted with twin 30 mm cannon. It first flew in August 1954 and was used for handling, performance and development work. Now on display at RAF Cosford. (BAE SYSTEMS)

Wing Commander Roland 'Bee' Beamont CBE, DSO & Bar, DFC & Bar, the famous chief test pilot for the English Electric Company, following a distinguished wartime service career. He performed an important role in the development of the Canberra, Lightning, and TSR2. (Ian Whalley)

A quartet of No. 111 Squadron F1A Lightnings, all built in 1961. The F1A differed from the F1 in being capable of in-flight refuelling. This mark was flown by the 'Treble One' until replaced by the F3. The aircraft furthest from the camera assumed decoy duties in Germany in 1965. (Rolls-Royce)

F3 XR716 of No. 5 Squadron at RAF Chivenor in 1987, two months before its transfer to RAF Cottesmore for battle-damage repair. It was scrapped in 1994. (Jonathan Francis)

Two Lightning F Mk3s take off on patrol. (BAE SYSTEMS)

T5 XS456, built in 1965, of the RAF Lightning Training Flight at RAF Binbrook in 1987 at her final air show, after which she was sold into private ownership. (Andy Chetwyn)

No. 23 Squadron's XR753 shadowing a Russian Tupolev Tu-95 'Bear' in the mid-1960s. Now preserved by No. 11 Squadron at RAF Coningsby. (BAE SYSTEMS)

The over-wing fuel tanks are clearly visible on this 1966-built Lightning F6 XS919. It was sold into private hands for static display in 1983. (BAE SYSTEMS)

XP693 Lightning F6 on static display at RAF Wattisham in July 1992. Its first flight was in 1962 as an F3 before conversion to F6 standard. Used entirely for tests and development work until 1992, after which it was sold to Thunder City in Cape Town and returned to flying condition as ZU-BEY. (Kerry Taylor)

This was the first T5 trainer produced. XS416 flew for the first time in August 1964, and is seen here landing at RNAS Yeovilton in July 1986. (Kevin Slade)

The first Lightnings delivered to the RSAF (Royal Saudi Air Force) were two F54s. 54-650, the one closest to the camera, is an ex-RAF XM989, while 54-651 is an ex-RAF XN992. Both have been preserved: XM989 is on display at Dhahran Airport and 54-608 at Prince Sultan Air Base Al Kharj, Saudi Arabia. (BAE SYSTEMS)

Lightning F.53 – 53-693, ahead of 53-684 in this image, was delivered to the RSAF in Jeddah in May 1969. Retired in 1985, it was finally sent to the East Midlands Aeropark. 53-684 was delivered in 1968 and crashed on take-off from Dhahran in June 1980. (BAE SYSTEMS)

V-bombers

Three technological developments in the 1940s rendered the famous wartime bombers obsolete: the advent of the jet engine, the nuclear bomb – and the threat it posed – and the electronic revolution, which allowed remote delivery of deadly weapons with great accuracy. In 1946 the Ministry of Supply issued a specification for a bomber capable of carrying a nuclear or conventional bomb load at 50,000 feet at close to supersonic speeds. Known as the V-bomber programme, it produced the Vickers Valiant, Handley Page Victor and Avro Vulcan. In April 1956 *Flight* magazine compared the embryonic V-force with the USAF Strategic Air Command and observed that 'the importance of the V-bomber, both as the carrier of frighteningly massive retaliation and as a national capital investment, cannot be overestimated'.

Vickers Valiant

Vickers tender for the new bomber was initially unsuccessful, but they persisted and an order for a prototype and production aircraft was placed in April 1948. In a show of extreme efficiency the first production aircraft flew on 21 December 1953, operational squadrons were equipped in 1955 and, by August 1957, the production run was complete. This was achieved despite the challenges faced by the designers and engineers at Vickers: increased air speeds, operating temperatures, required altitudes, navigation in these new conditions and the necessity for precision bombing. The prototype was powered by four Rolls-Royce Avon 200 series engines providing 10,000 lbs of thrust each, fed through leading edge intakes.

Across all three V-bombers was the rather curious and dangerous crew escape procedure. Because the pilots sat high up in the fuselage, it was possible to provide them with Martin-Baker Mk3 ejection seats. The solid canopies would be blown off and the pilots ejected. However, the three other crew members were seated side by side, facing rearwards deep below the pilots, and had to bail out the old-fashioned way. This was very different from exiting from a piston engine aircraft; the speed was greater, the slipstream an issue and everything happened faster. They had to jettison the main entrance door, extend the slipstream shield and attempt to exit. In addition, Vulcan crews had to avoid the nose gear if it was down; Victor crews exited almost level with the air intakes, whereas Valiant crews had no obstructions.

By the time the second Victor prototype emerged, underwing hard points had been provided on which to hang 1,000-gallon fuel tanks and, in 1954, in order to fill a gap in the RAF arsenal a reconnaissance/bomber version was developed to provide high-altitude photography.

The RAF had a considerable amount of preparation work to complete. Runways had to be extended, large hard standings built and completely new training programmes and documentation established.

By April 1956 three squadrons, Nos 138, 214 and 543, had been formed and a fourth was on the way. In the meantime, the Valiant B(PR)1, the dual-purpose reconnaissance and bomber version was officially announced. This had extra fuel tanks in the forward part of the bomb bay and, further back, a pannier with cameras; in order to accommodate these, the bomb bay doors had sliding windows.

This was the same year that the Valiant went to war. As part of Operation Musketeer, twenty-four Valiants were relocated to Malta and on 31 October 1956 they began a two-day bombing attack against Egyptian airfields; the first large jets to drop bombs in combat. Part of a combined French and British force, they attacked Egypt in an attempt to reverse the Egyptian government's nationalisation of the Suez Canal.

Designed as a deterrent force, the Valiant had to be capable of delivering nuclear weapons. From 1955 a number of bombs were available but no live tests had been completed. In October 1956 with the dropping of a nuclear bomb on a range at Maralinga, Australia, the airborne deterrent became viable. To counter the dangers of a nuclear flash all V-bombers were painted in anti-flash white in an attempt to deflect some of the thermal radiation and protect the crew and aircraft.

By the mid-1950s Britain had fallen behind the United States and the Soviet Union in nuclear technology. In an effort to catch up they undertook Operation Grapple, which was created to test a hydrogen bomb. Malden Island was selected as the target, with flights and control managed from Christmas Island. In May 1957 XD818 dropped an H-bomb from 45,000 feet and Britain became the first country to air drop such a weapon. Although not fully successful, the British scientists overcame technical difficulties to prove the concept, but in the end an American weapon was selected.

After its pioneering role in nuclear warfare, the Valiant was probably best known for its valuable contribution as an inflight fuel tanker. Flight Refuelling Ltd produced a probe and hose-reel pack that allowed the Valiant to be refuelled in flight and, in turn, become a tanker. Although not ideal as a tanker and soon replaced by the Victor with multiple refuelling points, it had, however, established in-flight refuelling as an essential element of RAF operations.

Once again technological advancements in the Jet Age resulted in what was a fine aeroplane becoming obsolete after only a short service life. The V-bombers were designed to fly at high altitudes and, when it became clear that missile technology could reach these heights, any aircraft that was not fitted with ECMs would be considered non-operational. The Valiant force swopped its anti-flash white for the green and grey camouflage of low-level tactical bombers.

New alloys had been created to meet the demands of jet flight and the chemical reactions between these were relatively unknown. When cracks were found in a Vickers Viscount, it followed that the Valiant, using the same alloy, would suffer similar problems. Cracks were found, but the issue came to a head when, while in flight, the crew of one aircraft heard a wing spar crack. All Valiants were grounded and repairs were trialled on one aircraft but they proved too expensive and the fleet was retired in January 1965.

Handley Page Victor

All three V-bomber manufacturers approached wing design differently: Vickers used a moderate sweep, Avro the delta and Handley Page the novel crescent shape. This was a wing that was swept back 53 degrees at the wing root, then 35 degrees, terminating at the wingtips at 22 degrees. The principle was to smooth the changes in the sweep and maintain a constant speed along the full wing in order to overcome the compressibility problem experienced as an aircraft neared Mach 1.

The first flight of the HP80 – prior to it being named the Victor – was on Christmas Eve 1952. This aircraft with its crew of four was lost in July 1954 when the tail broke away in a low-level flypast. Tail plane attachments were redesigned and the second prototype flew in September. By February 1956 the first production aircraft were completed. These were fitted with four Armstrong Siddeley Sapphires, each providing 11,000-lb thrust. Again a first was achieved by a British jet when a Victor achieved Mach 1.02 in a shallow dive in 1957, the largest aircraft to 'break the sound barrier'.

XA931 was the first BMk1 to go into service with No. 232 Operational Conversion Unit (OCU) in 28 November 1957. Unusually, a basic flight simulator was delivered to the OCU. By the following April the Victor was in frontline operations with No. 10 Squadron. It had the ability to carry more bombs than the Vulcan and the flexibility to carry a wide range of weapons from the Blue Danube nuclear weapon to Tallboys and 48,100-lb bombs. Victors were first used in active operations during the Malaysian Confrontation when they were stationed at Tengah and RAAF Butterworth. Although at one stage they were on immediate standby for three days, they were never actually used against the insurgents; nevertheless, their mere presence was an important deterrent to the Malaysian forces and anyone else who thought to enter on their side.

Continued development in missile and air defence technology gave rise to the BMk1A, which was fitted with early ECMs. They also forced Handley Page to develop a more powerful Victor that could reach an altitude safe from the Soviet surface-to-air missiles (SAMs). Re-engined with four 17,250-lb-thrust Rolls-Royce Conways, the B2 had a larger wing and slightly stretched fuselage.

The Blue Danube was replaced by versions of the American-supplied Yellow Sun H-bomb, but both of these weapons were free-fall and required the delivering aircraft to be over the target. Defence developments along with countering attack technology required that the RAF could deliver self-propelled stand-off weapons to the selected targets. The British-built Blue Steel met these criteria, but the Victor had been designed for the conventional bombing role. Apart from not having the necessary on-board controls and management, the ground clearance of the Victor made loading the Blue Steel difficult; in fact, folding fins had to be fitted and then there was only a 14-inch clearance to the runway once in place. Bomb doors were removed and replaced by fairings that fitted around the missile.

Put into service by February 1962, twenty-one of the thirty-four B2s were converted to BMk2R standard with Küchemann carrots, fairings on the wing trailing edge designed to delay boundary layer separation at height when armed with a Blue Steel. In addition, they could carry underwings tanks, had a rapid-start system and possessed more powerful Conway engines.

An additional nine B2s were modified for the reconnaissance role as B(SR)Mk2s to fill the gap left by the withdrawal of the Valiants. With a 40 per cent greater range and operating at a higher altitude, the Victors were outstanding in this role. With up to eight cameras mounted in a fan pattern, they could cover an area the size of the United Kingdom in two hours and the whole North Atlantic within four hours. When using radar mapping, they could plot every maritime vessel within range. These aircraft were also used to 'sniff' the radiation emitted by French and Chinese nuclear tests.

Soviet SAMs forced the RAF to change tactics and the Victors were switched over to a low-level attack role from March 1963. They were camouflaged against high-level Soviet fighter aircraft, which at the time had no 'look-down' radar. A great deal of experimentation and trials were carried out to determine the most effective way of launching low-level attacks. The Victor squadrons served in the QRA role with aircraft on a rolling permanent readiness, but this all ended when Blue Steel was replaced by Polaris, ending the Victor's part in the UK's nuclear deterrent strategy.

Air refuelling capability proved to be vital in the Jet Age. Jets were thirsty and they could not reach their expected targets, even with full tanks. By 1962 Victors had been identified as candidates to become tankers and plans were made to convert the older BMk1s. Industrial disputes and the upheaval caused by government attempts to merge British aviation companies into a single organisation lead to a two-year delay in the process. The withdrawal of all Valiants in 1964 created an immediate need and, as an interim solution, a small number of Victors were to be fitted as two-point tankers. These emergency conversions were carried out separately, so as not to disrupt the main conversion programme. The two-pointers were delivered in October 1965. The main conversion contract delivered the three-point tankers; designated KMk1 or KMk1As, indicating the second batch of converted Mk1 airframes. These tankers operated until the mid-1970s.

With greater power the BMk2 was a better base on which to build a tanker. The KMk2 could take off in hot and high conditions with a full load and was able to reach the fighters' altitude. Twenty-four of these tankers were produced, which amazingly could carry more than their own weight in fuel.

The Marham-based aircraft of Nos 55 and 57 Squadrons were heavily used, year after year, supporting air defence Lightnings and Phantoms, Vulcan operations and Strike Command Buccaneers and Jaguars. They served a vital role in the Falklands War both as tankers and in the maritime radar reconnaissance role. Before the Nimrod took over, Victors had flown round-trip missions of nearly 15 hours to the Falklands and then spent 90 minutes scanning the islands and surroundings, mapping Argentinian land and sea activity. Afterwards, they would refuel the Nimrods on their long reconnaissance missions. The most famous bombing raids by the modern RAF were the seven Black Buck Vulcan attacks on the Falklands. For these, typically eleven Victors and two Vulcans were launched, which included a spare aircraft of each type. These well-documented attacks had a very complex pattern of leap-frog refuelling the Victors and Vulcans in order to give the single selected Vulcan the endurance to fly a non-stop return distance of 6,800 nautical miles, at the time the greatest range ever flown on a bombing mission.

To refuel the Lockheed Hercules, which carried out Falkland supply drops, the Victors overtook the slower-flying Hercules, which then connected, and the two aircraft entered a shallow dive as the top speed of the transporter was lower than the lowest speed of the tanker. Once the airfield at Stanley was repaired in May 1985, the Hercules could land and the need for refuelling over the South Atlantic Bridge was over.

High utilisation during the 1982 war had used up airframe hours and the Victors began to be retired from 1985. When Kuwait was invaded in 1990, eight Victor KMk2s of No. 55 Squadron were the first RAF tanker force to be deployed. They were then supplemented with nine VC10s and two Lockheed Tristars. The Gulf War marked the swansong of the Handley Page Victor and they were withdrawn from service in October 1993.

Avro Vulcan

The late 1940s and early 1950s were an era of innovation and experimentation. The famous designer of the Lancaster bomber, Roy Chadwick, proposed building a delta-winged aircraft to meet the 1947 specification for a four-engined nuclear bomber. Initially merely a flying wing, the design evolved to include the more conventional tail fin and extended fuselage, which became so recognisable as the Avro Vulcan.

Roy Chadwick was killed in a flying accident in 1947 and Stuart Davies replaced him as head designer at Avro, continuing with the development of what was then known as the Type 698. Both Avro and the Ministry of Supply wished to confirm the feasibility of the delta design and the proposed flight controls. Funding was received to build a series of scaled versions of the larger bomber. These Type 707 aircraft were purely experimental but were necessary to ensure funds were not wasted on the full-scale bomber.

The first Vulcan VX770 flew in August 1952. Powered by lower-rated Sapphire engines, and fitted with a single ejector seat and a temporary fuel tank in the bomb bay, the aircraft made its dramatic appearance at the Farnborough Air Show. Without the benefit of computer systems and simulations, test pilots had a great deal of influence on how an aircraft evolved. Roly Falk, the chief test pilot, argued that, as the controls were powered, there was no need for the traditional 'spectacle' type control wheel and proposed fitting a fighter 'joystick' instead. The Vulcan became the first large aircraft flown in such a way; this was only seen in a large jet from the mid-1980s, when Airbus introduced the 320 range controlled via a side stick, very much like a computer joystick.

The first few aircraft flew with straight leading edges to the wings but suffered from buffeting as the airspeed reached near Mach 1. A new wing, known as the 'Phase 2' wing, was designed with 10 degrees less sweep in first section, while the outer section retained the 52 degree sweep.

Delivery to the RAF of the first B1s began in May 1956, with No. 83 Squadron being the first operational unit. The following year saw the release of the infamous Defence White Paper, which, apart from cancelling many projects, also served to restructure the defence organisation. In order to be a credible deterrent, the Vulcan required ECMs; these were installed in a new larger tail cone, which held the Red Steer early-warning radar.

Early Vulcans were soon superseded by a more powerful, enhanced version. The B2 first flew in August 1958 and it featured more powerful Olympus engines, could be refuelled in flight and was fitted with a new wing, the Phase 2C, 519 square feet larger than those fitted to the B1A and providing a 25 per cent increase in range. These went into service in December 1960.

Initially, the Vulcan was tasked with delivering Britain's first nuclear weapon – the free-fall Blue Danube. It was designed to be dropped in a conventional manner like the bombs dropped by Lancasters. This became increasingly dangerous with the advent of SAMs. The Blue Danube was replaced by the stand-off weapon Blue Steel rocket-powered bomb, which could be launched 100 miles from the target. In October 1962, while the missile was under test in Australia, the decision was made to switch the Vulcan's to low-level delivery in order to counter the high-altitude Soviet defences.

However, the standoff range was very short and Britain looked to the United States for a solution. The Skybolt missile was offered with an advertised range of 1,200 nautical miles, but on the horizon was the Polaris submarine-launched missile. Using submarines provided far more flexibility and also the power of surprise; as a result the US government cancelled the Skybolt programme.

The Royal Navy purchased Polaris and took on the nuclear deterrent role. Blue Steel was also retired and the Vulcan became a tactical nuclear strike craft carrying the WE177B nuclear bomb. Up to twenty-one conventional 1,000-lb bombs could be delivered and crews carried out training to maintain their currency, which paid dividends in the 1982 Falklands War. Virtually at the end of its service life, the Vulcan went to war in the most spectacular fashion. Refitted with refuelling probes that had previously been blanked off, five old Vulcans flew to Ascension Island. The famous Black Buck raids resulted in minimal physical damage, but they re-established Britain's ownership of the islands and threatened possible attacks on the Argentinian mainland.

Demand for tankers during the Falklands War, and the fact that the conversion of additional VC10s and Tristars would take twelve months, gave rise to the Vulcan K2 tanker. Rapid conversion of six aircraft comprised fitting three large tanks in the bomb bay, replacing ECM equipment with the hoses and drum and fabricating a rather square box under the tail cone, which held the drogue unit. Their role was short lived and, within two years of the Falklands War, the last Vulcans were retired from active service.

The first flight of the Valiant prototype on 18 May 1951 lasted 5 minutes. Finished in polished aluminium, WB210 carried out flight testing until January 1952, when it caught alight while in flight. It crashed near Hurn with the loss of the co-pilot Squadron Leader Foster, who ejected but hit the tail. (Rolls-Royce)

Another picture of XD816 topping up Lightning F1A XM179 from No. 56 Squadron in 1960. (BAE SYSTEMS)

When switched to a low-level role, the upper surfaces were camouflaged, as seen on this Valiant. (BAE SYSTEMS)

XD816 B(K)1 flew three missions over Egypt during the Suez crisis of 1956. In 1964 the airframe was used by Vickers/BAC for trials and modifications of the wingspar, as seen here in April 1968, when it was on probably one of its last flights. It flew into Abingdon for the RAF's fiftieth anniversary in 1968, the last Valiant to fly, and was struck off charge in 1970. The nose section was removed and the aircraft is now preserved at The Brooklands Museum, 29-04-1968 WCN2631. (BAE SYSTEMS)

Handley-Page HP.80 Victor B.2, most probably XH669, at Farnborough in 1960 for the first time. It was converted to a K2 tanker in 1970 and served until 1990, when it was damaged by a hot-air leak. The cockpit is currently being restored. (John Kendall)

A dramatic display by a Handley-Page HP.80 Victor B2 at the 1960 Farnborough Show. The show was opened every day by a V-bomber scramble from the western end of the field. (John Kendall)

A line of new Victor B1s at Handley Page's Radlett airfield in the summer of 1958. (BAE SYSTEMS)

Victor K1A XH650
was an interim tanker,
which could refuel
only two aircraft at
a time, as seen here
with Lightning FMk3s
XP764 and XP707 of
No. 29 Squadron. (BAE
SYSTEMS)

Converted to a tanker
in September 1964,
this No. 57 Squadron
Victor BK1 XA926
clearly shows the
crescent wing shape
and the guide mark that
receiving aircraft used
to line up to the drogue.
(George Woods)

XL189, a Victor K2,
with all three refuelling
drogues trailing and
ready for business. (BAE
SYSTEMS)

A Victor K2 tanker at the end of its landing run with the braking parachute fully deployed. (BAE SYSTEMS)

The prototype Vulcan B1 VX770, which broke up during a display at RAF Syerston in 1958 when, at low level and high speed, the pilot began a rolling climb and the starboard wing disintegrated. (BAE SYSTEMS)

Three Vulcan B1s, XH476, XA909 and XH475, in the late 1950s or early 1960s. (BAE SYSTEMS)

Delivered as a Vulcan B1 in 1957, XA911, pictured here in 1965, was converted to B1A standard in 1962. (Niel Aird)

Avro Vulcan B2 XH534 at Farnborough in September 1960. The second MK2, it was fitted with production Olympus 201 engines and an ECM tail cone. It was used for trials until 1964, when it was allocated to No. 230 OCU. In 1973/74 it was converted to a B2 MRR (maritime radar reconnaissance) aircraft and flew with No. 27 Squadron until scrapped in 1981. (John Kendall)

Vulcan B2 XM575, built in 1963, was the second aircraft to be fitted with upgraded Olympus 301 engines. Initially a Blue Steel aircraft, it was converted to the conventional bombing role. Now preserved at East Midlands Airport. (BAE SYSTEMS)

Solway Aviation Museum's Vulcan B2A XJ823 flight deck, clearly showing the fighter-type controls. (Peter Clarke)

XJ823's lower cockpit, where the navigator radar, navigator plotter and air electronics officer worked facing rearwards. This Vulcan is on display at www.solway-aviation-museum.co.uk. (Peter Clarke)

1960 Avro Vulcan K.2 XH560 of No. 50 Squadron lifts off from RAF Wyton in 1983. The wide white strip and red markings were used by the pilots of refuelling aircraft to line up on the drogue. (Andy Chetwyn)

Vulcan XH560 was converted to a maritime reconnaissance role and then to a K2 tanker; it is pictured here refuelling a Lightning F6 of No. 11 Squadron. (BAE SYSTEMS)

Vulcan K2 tankers buddy refuelling. Here XH560 is restocking from XH561. (BAE SYSTEMS)

Vulcan XM575 over Lincoln Cathedral. (BAE SYSTEMS)

Airliners

Britain was a pioneer in jet-powered aircraft and not the least in jet airliners. The De Havilland Comet had been the first commercial passenger jet, breaking rules and records. The second generation were no different, and the pilots of the time followed in the footsteps of their elders, experimenting and learning at every opportunity.

During their training British European Airways (BEA) pilots undertook a 'High Flight' where, after hours of classroom lessons and simulator flights, they learnt boundaries and technique. First, with a high-speed run up to Mach 0.92, was recovery without pitching the nose down or up beyond retrieval, by not touching the speed brakes and instead only slowly pulling the throttles back. Next was the Dutch roll, which was kept in check by yaw dampers; however, when they were switched off, 'very soon the aircraft start[ed] rolling left then right and while it [was] rolling it [was] also yawing' [Andrews]. The pilot would 'wait till it's in about mid-swing and apply a chunk of opposite aileron briskly and remove it equally briskly' [Andrews]. Finally was the experience of approaching a stall in a large jet airliner, feeling the warnings, and reacting before it was too late.

BAC1-11or One Eleven

In an effort to save the United Kingdom's failing aircraft industry, the government forced a merger of Vickers, Hunting, Bristol and English Electric. Vickers had been working on a smaller version of their VC10, while Hunting had been working on a thirty-seat commuter jet. The design ideas were combined and then refined to produce an eighty-seat aircraft for the general market. It was the first jetliner not built to a specific user's requirements, which provided it with a broader market.

Only twenty-seven months after the go-ahead to build the BAC1-11 was taken, the prototype took to the air from Hurn on 20 August 1963, accompanied by a Jet Provost chase plane. *Flight* magazine commented that, 'rarely has a new aircraft taken to the air backed with so much sound commercial hope'.

It was watched by several thousand BAC employees, future operators and hundreds of onlookers. Jock Bryce, BAC chief test pilot, and Mike Lithgow, co-pilot, who had completed a number of ground tests in inclement weather, decided the aircraft was ready to fly. Wearing the colours of the first customer British United Airways (BUA), it lifted off after a run of 1,100 yards and flew for 28 minutes.

It was also the first aircraft on which the 'stick shaker' was introduced. Within two months of its first flight, the prototype was destroyed in a crash with the tragic loss of its renowned test pilot, Mike Lithgow, and six members of the test team. They had encountered an unrecoverable 'deep stall' where, at a certain angle, the airflow over the T-tail-mounted elevators and rudder was masked by the main wings and the aircraft simply dropped out of the sky.

It was a huge blow to the new company. The leading edges of the wings were redesigned and powered elevators were developed but, most significantly, BAC developed an anti-stall warning system known as the 'stick shaker'. This shakes the control column several knots prior to the occurrence of a stall and the pilot can then prevent a deep stall. In the words of Roger Andrews,

> Since there is no natural stall warning the CAA have mandated a warning system. This shakes the stick with such violence that a pneumatic drill operative would be alarmed. If you ignore the shaker, next comes the pusher. This is no namby pamby nudger as fitted to some transport aircraft, this is a full blown pusher. Sixty pounds of force throw the stick forwards, out of your grasp and holds it fully forward (not just neutral or a bit forward but fully forward) until the airspeed returns to a sane figure.

Prior to the tragic accident a number of orders had been placed, many from America where it was cheaper and more advanced than its rival the Douglas DC9. The first production aircraft, a Series 200 model, was handed over to BUA in January 1965. Only nine of the heavier 300 Series, with additional seating and heavier engines, were built before American legislation forced the development of the 400 Series in mid-July 1965. American Airlines bought thirty of the 400 Series, which made them the largest BAC1-11 customer. As Peter Lewis remembers:

> It was a lovely aeroplane to fly. You took off, cleaned up and accelerated to 333 knots and cruised. A skill was developed among the flight crew as to the exact point out from touch down to close the throttles and begin the descent. The aircraft would then descend to 1,000 feet, where the pilots would prepare for finals; it was all about trim and accuracy.

Demand from BEA for greater seating capacity resulted in the 500 Series stretched BAC1-11 with 119 seats. Traditional delays within the British aircraft industry meant that the DC9 had caught up and surpassed the BAC1-11 in development, and there were to be no more sales of the BAC product into the United States. The BEA aircraft had cockpits redesigned to provide commonalty with the airline's Hawker Siddeley Tridents.

The best performer of the BAC1-11 family was the last British-built model, the 476 Series, which mated the longer wings of the 500 with the body of the 200 and was aimed at the Third World market. It had a modified undercarriage to be able to land on rough-and-ready airstrips and two examples were fitted with large side cargo doors. Marketed as a competitor to the newer Fokker Fellowship, it was not, however, a commercial success.

When the British production line closed, all the tooling and jigs were bought by the Romanian company Romaero SA. The company planned to build an additional twenty-two aircraft; however, production of each unit took longer than expected and only nine ROMBAC1-11 561Cs were completed.

A total of 244 aircraft were built and a number remained in service in Europe until the noise-abatement laws led to their replacement in the 1990s. Subsequently, they were purchased for operations in the Far East and particularly in Africa, where a Nigerian airline was the largest user, while a number were used on scheduled routes in southern Africa.

The last UK flight was on 26 April 2013 when ZH763 was retired from testing work with QinetiQ and flown to Newquay for preservation. Two BAC1-11s still fly as test aircraft with the American aviation company Northrop Grumman.

Hawker Siddeley Trident

Starting off as the De Havilland DH121, the Trident was developed during an upheaval in the British aviation industry, which delayed its entry into service and gave the lead to its rival, the Boeing 727. Originally designed as a larger aircraft with high passenger capacity and powerful engines, BEA persuaded De Havilland to reduce the size of its design to meet the airline company's projected future overcapacity. Tailoring the product to the 1956 BEA specification for a second-generation jet to serve its European market impacted on its ultimate commercial success. It was developed for short intercity flights within Europe and did not easily fit other routes.

For safety, the world's first tri-jet airliner had each of its engines driving its own hydraulic system, with number one engine driving the green system, number two the yellow and number three the blue. The additional costs of three engines put the Trident at a disadvantage when compared to the second-generation twin-engine airliners. High-lift wing design, which was the most sweptback of any airliner, coupled with more powerful Rolls-Royce Spey 511 motors, was developed for the BAC1-11; this provided higher operating weights and capacity, which was competitive even with the Boeing 727. It became one of the fastest subsonic airliners and would cruise at more than 610 mph or Mach 0.84. To achieve these results, the wing provided less lift at lower speeds, requiring a lengthy take-off run, which led the aircrews to nickname the Trident 'The Ground Gripper'.

Development that had taken place in an attempt to solidify American Airlines' interest was incorporated into the Trident 1C. This mark first flew in 1962 and entered BEA services in 1964. It soon became evident that BEA had been mistaken and the original design was more suited to the changing airline transport market.

The Trident Two, when branded the 1E, had been fitted with the Spey 511; however, the 2E for BEA was fitted with the Spey 512 and produced an additional 8,500 lbs of thrust. It also had an improved wing, which increased lift, while a greater capacity and range was provided by the integral fuel tank in the fin. Uniquely, it was possible to engage reverse thrust on the outer two engines while in flight and experience a tremendous rate of descent. Peter Lewis recalls a flight from London to Gibraltar:

Throughout the flight ATC (Air Traffic Control) had been requesting we call the company but we were out of range. As we approached Madrid the local ATC relayed a warning that there was a report of a bomb on board. Captain Johnny Truscott requested clearance to land at Madrid, which was directly below. He was cleared to land on whatever runway was best for him and he engaged the reverse thrusters. The Trident descended from 33,000 feet at up to 15,000 feet a min and we landed downwind on the non-duty runaway for a very quick evacuation and search that only turned up some explosive Hilti-type refills.

It was on Thursday 10 June 1965 that the passengers on flight BE343 aboard BEA Trident 1 G-ARPR, travelling from Paris to London, made aviation history: the first commercial auto-flare landing. This was the first stage of fully automatic landings that now occur daily hundreds of times worldwide. The duplex Smiths Autoflare automatic blind-landing system had been fully certified for use in the Trident, the first aircraft designed from the outset to take automatic landing equipment. The system was eventually fitted in triplicate in the Trident 2E, with all three operating simultaneously and constantly comparing their findings. Interestingly, in the event of discrepancy, a majority vote puts the third channel out of operation. Any subsequent failure saw the system hand over a trimmed aircraft to the flight crew. As 'AJ' says:

> Trident's auto-land capability was revolutionary, though the mechanism used to achieve this seems quaint nowadays, employing triplex parallel circuits from ILS receiver to control surface actuator. But remember, this was an electro-mechanical system, no electronics.

A forerunner to the ubiquitous 'black box' was another Trident first, as 'AJ' continues:

> In order to gain approval for its ground breaking Cat3 auto-land, all the aircraft were equipped with a 72-channel flight data recorder, extraordinary at the time.

BEA began to explore the availability of larger aircraft and considered the Boeing 727, but the government required that they support the British aircraft industry. Hawker Siddeley immediately proposed a stretched fuselage version of the Trident, to be known as the 3B, which first flew in December 1969. This model could carry 180 passengers in high-density configuration and above the centre engine was a small Rolls-Royce RB162 booster unit providing an additional 6,000-lb thrust on take-off – the first aircraft to use the booster application in commercial service.

In August 1977 almost half of the British Airways' Trident 3 fleet was grounded when cracks in the wings were found. G-AWZL, which only had 8,600 flights and 10,929 flying hours, experienced a persistent fuel leak and this led investigators to find the cracks. Repairs were made to all the aircraft, which returned to service. It showed that there was still a great deal to learn about metallurgy and fast jets.

Foreign orders

Hawker Siddeley embarked on an extensive worldwide sales tour, borrowing six BEA aircraft for the promotion. The competitive environment was made more so by a simultaneous

Boeing 727 tour. At one point a Trident and 727 found themselves parked together at Karachi, where the two crews examined one another's product. It appeared that the Americans were responsible for a brochure that compared the Trident unfavourably to the 727, to which De Havilland at the time responded appropriately.

Small numbers were sold. The first three went to Kuwait Airways in November 1964 and three to Iraqi Airways, followed by four to Pakistan International, while others were leased to foreign airlines. The Pakistan 1E aircraft were sold to China and this resulted in the largest foreign order for the type. The Civil Aviation Administration of China purchased six 2E aircraft in August 1971, followed a year later by another six and a third set of six in November 1972, together with two Super 3B variants, which had an extended range.

The final order was in December 1973 for fifteen Trident 2Es, which brought the total sales of Tridents to the Chinese to thirty-five aircraft. Most of these aircraft eventually moved from the national airline and served in the air force, some flying up until 1997.

When production ended in 1978, 117 Tridents had been produced, in comparison to 1,832 Boeing 727s. British Airways began retiring the aircraft in 1978, with the final 3B being withdrawn in 1985.

Vickers VC10

The indecisiveness and poor management demonstrated by the British Overseas Airways Corporation (BOAC) executive and the government saw promising airliner developments delayed and cancelled. The VC10 would have been on the market far sooner, and therefore able to compete effectively with the Boeing 707 and Douglas DC8, if the go ahead to develop it had been given two years earlier.

BOAC signed a contract for the supply of thirty-five VC10s in January 1958. In June 1960 they ordered ten Super VC10s, designed to seat 212 passengers but, before this much larger aircraft was designed, BOAC changed their orders. Fifteen standard VC10s had already been made when the sixteenth was built as a much-reduced Super VC10. Instead of having the fuselage stretched by 27 feet, the revised Super VC10 was only 13 feet longer, providing seating for 175 passengers and powered by Conway 43 engines producing 22,500 lbs of thrust each.

The link between the contractor and the airline was unhealthily close, to the point where BOAC had ultimate technical control of development. Sir George Edwards, managing director of BAC, said: 'In our view – and this view is shared by BOAC – there will be a continuing world need for a large and economical subsonic jet for many years to come.' Accordingly, the corporation issued a specification in 1956 for a big jet to fly the company's southern and eastern routes. It was important that the design had good airfield performance compared to other big jets that required longer and longer runways. The rear engines had a weight penalty that the engineers argued was countered by the clean high-lift wing and the ease with which more powerful engines could be mounted and the fuselage stretched.

After two months of ground tests the first VC10 G-ARTA flew on 29 June 1962. It was to be crewed by two pilots, an engineer and a navigator. This was BOAC's large 'second-generation' jet airliner, which could operate from airfields on average 2,000 to 3,000 feet smaller than those required by the Boeing 707 or Douglas DC8. When it entered service on 1 April 1965, the Super VC10 was the largest British transport aircraft ever built.

Within a year the VC10 was proving very reliable and popular. Where other aircraft required six years of service to establish extended overhaul intervals, the VC10 achieved this within twelve months, when the hours were extended from 4,000 to 6,000, which meant two hours less engineering time per flying hour compared to the Boeing 707-420. Pilots easily converted from piston aircraft, saving thousands on training costs, while passengers enjoyed the quiet cabin.

The aircraft, however, never managed to break into the mainstream market, possibly because it did not show any major economic advantage over the larger proven American products. With the introduction of the larger Boeing 747 in the early 1970s, the VC10 was no longer economically viable and they were all withdrawn by March 1981.

Apart from BOAC, Ghana Airways ordered three, one of which was cancelled, and BUA ordered two standard VC10s. The Ghanaian aircraft were fitted with a cargo door, while the BUA models were built as combi versions, providing mixed passenger and freight abilities.

The RAF placed an order for fourteen VC10 C1s with the more powerful engines, but with the standard fuselage, in order to serve in a dual role as a passenger and freight transporter, with a crew of nine and the capacity to carry 124 passengers. The large side cargo door allowed the aircraft to be converted into a passenger/freight combination or a full-freight aircraft. In casevac configuration it could be fitted with up to sixty-eight stretchers.

In 1993 these airframes were converted to provide air tanking, with the fitting of a refuelling pod under each wing, allowing two aircraft to be refuelled at a time from the 69,800 kg of fuel held on board. They were also fitted with a refuelling probe so that they could in turn be refuelled themselves.

Inflight refuelling is a basic necessity for jet-equipped air forces and, by the late 1970s, tankers were needed to support the imminent Tornado fleet. The current tankers were converted Victor and Vulcan bombers, which were reaching the end of their service life. After a feasibility study, it was decided to convert ex-civilian VC10 airliners to tankers. These aircraft required considerable engineering work but the time and expense paid off as they served the RAF for nearly twenty years.

Nine airframes, ex-East African Airways (EAA) and Gulf Air were immediately available from British Aerospace (BAe). These were purchased and converted by BAe at Filton in Bristol between 1980 and 1987. They were fitted with three refuelling points – EAA aircraft with uprated Conways – while the cockpits and systems were modified to ensure commonality with the C Mk1 fleet.

The ex-BOAC/Gulf Air aircraft that became K2s and the EAA-aircraft-designated K3s were completely stripped down and rebuilt. The K3s had the advantage of fitted cargo doors, while the K2s had to have a hole cut in the roof to fit the five fuel cells inside the fuselage. All were fitted with rear-mounted hose drums and new units designed to house the refuelling point under the tail. Above that point in the extreme tail, auxiliary power units were fitted to provide self-sufficiency on un-serviced airfields. Tanker-to-tanker refuelling was provided through a refuelling probe installed in the nose.

When in 1981 British Airways retired the VC10s, fourteen were purchased by the RAF to replace the older K2 tankers. The airframes were stored for nine years at RAF Abingdon until, in 1990, BAe were contracted to convert five of them to K4 tankers and to fit refuelling nose probes to the C Mk1, making them C Mk1Ks. A seven-year contract involved the complete rebuild of each airframe and the delivery of virtually new aircraft, as years of standing had resulted in deterioration, extensive corrosion and out-of-date certifications; they were not airworthy.

A low-infrared paint scheme was applied to the fleet, along with radar and infrared counter-measures to provide some self-defence capability. The last VC10 flight took place on 25 September 2013, when ZA147 landed at Bruntingthorpe to be scrapped. Ollie Alderson reminisces:

> I loved the VC10, it was a pretty fast aircraft cruising at up to Mach 0.88 and during air testing we flew it at Mach 0.925.

The prototype G-ASHG in flight. It crashed on 22 October 1963, near Chicklade in Wiltshire, with the loss of all on board. (BAE SYSTEMS)

One-Eleven 201AC – C/n 011 G-ASJG, delivered to BUA in July 1965 – on the apron with BUA's first VC10, G-ASIW. (BAE SYSTEMS)

An early One-Eleven on show with typical 1960s brown interior. (BAE SYSTEMS)

One-Eleven 500s under construction in September 1968: G-AVML and G-AVMM ended their days derelict in Nigeria. G-AVMN is at the Panzer Museum in Fladholte, Denmark, while G-AVMO is preserved at the Scottish Museum of Flight; G-AVMP was scrapped in 2002. (BAE SYSTEMS)

G-AVMH was the first production One-Eleven 500 for BEA, delivered 10 weeks ahead of schedule in 1968. (BAE SYSTEMS)

British Airways BAC 1-11 Series 510ED, G-AVMJ, originally built for BEA in 1968, on the ramp at Heathrow in 1982 in the company of five Tridents and a Boeing 737-200. It now serves as a cabin trainer at Tresham College in Kettering. (Andy Chetwyn)

British Caledonian BAC 1-11-501EX, G-AXJM at Manchester Airport in 1985. (Jonathan Francis)

British Airways BAC 1-11-401AK, G-BBMF, at a damp Manchester Airport in 1988, dwarfed by the Boeing 757s. (Jonathan Francis)

G-AXMF was delivered new to Court Line in 1969, but was subsequently owned by companies in South America, returning to Europe in 1996. (BAE SYSTEMS)

ROMBAC 1-11 srs 561RC G-BNIH, *The Spirit of Europe* of London European Airways, parked at Luton Airport in 1988. Built for TAROM in 1986, she is now stored at Karachi. (Andy Chetwyn)

One of only two
One-Eleven 475s built
for Air Pacific of Fiji
in 1973. It was used on
its delivery flight for
sales demonstrations
to various airlines.
Seen here at Brize
Norton in 2007. The
forward fuselage has
been preserved at the
Bournemouth Aviation
Museum. (Jonathan
Francis)

After an accident
involving a One-Eleven
in 2002, the Nigerian
Goverment banned
all aircraft older than
twenty-two years. In this
2007 image, Okada's
large fleet is shown in
a poor state. (Akiwale
Makinde)

Trident 1C aircraft being
assembled for BEA at the
old De Havilland site at
Hatfield in Hertfordshire,
as part of Hawker
Siddeley Aviation. (BAE
SYSTEMS)

First off the production line, Trident 1C's G-ARPA and G-ARPB are seen here brand new in 1962. (BAE SYSTEMS)

Trident 1C G-ARPR at Heathrow in 1965, a month after performing the world's first fully automatic landing with revenue passengers as flight BE343 from Paris to London. (David Russon)

Delivered new in 1970 to Cyprus Airways as 5B-DAB, this Trident 2E has become famous with its sister ship 5B-DAE as the airliners shelled and damaged at Nicosia Airport Cyprus during the 1974 Turkish invasion. Marooned in the neutral zone, they are rotting away slowly. (Malcolm Nason)

Trident 2E BEA G-AVFC in her element. Built in 1967, she was broken up in 1981. (BAE SYSTEMS)

Under construction for CAAC of China at Hatfield, Trident 2E 2162 was delivered in 1973 registered as B2212. (George Woods)

Trident 2E B-2201 of the Civil Aviation Administration of China (CAAC) on a test flight with temporary British registration G-AZFT before delivery in 1972. (BAE SYSTEMS)

No. 2301 G-AWYZ was the first Trident 3B built, and is seen here in the characteristic steep climb of jet airliners. (BAE SYSTEMS)

On finals from the flight deck of a Trident 3B of BEA. (BAE SYSTEMS)

A busy Heathrow with a collection of fourteen Tridents, headed by a Trident 1c G-ARPM and followed by sundry Trident 3bs and 2Es. (BAE SYSTEMS)

First VC10 to cross the equator: G-ARVB c/n 805 is seen here over Johannesburg, South Africa, probably on tropical trails in 1963. (BAE SYSTEMS)

The Weybridge production line in late 1963. G-ARVK was delivered to BOAC in 1964, then to Gulf Air in 1975, and finally to the RAF as a K2 tanker. (BAE SYSTEMS)

Super VC10 G-ASGD in its element: from 1965 to 1980, it served with BOAC and British Airways until sold to the RAF. Registered as ZD232, it was stripped of parts to support the RAF VC10 fleet. (BAE SYSTEMS)

G-ASGD in New York among contemporary airliners: Boeing 707-465 G-ARWE, Bristol Britannia 312 G-AOVL and (in the background) National DC8 and TWA Boeing 727, 707 and (nearest) a Convair. (BAE SYSTEMS)

VC10 being prepared for flight: from left to right, a Commer Karrier ground power unit, AEC tanker, Hillman Husky van and an Austin J2 van. (BAE SYSTEMS)

The first Super VC10 G-ASGA, built in 1964 and seen here more than a decade later in Manchester after the creation of British Airways. Used as a K2 tanker from 1991 to 1994, as ZD230. (Jonathan Francis)

The first of five Super VC10s delivered to East African Airways, 5X-UVA was one of two VC10s to be destroyed in fatal accidents when overrunning the runway after colliding with debris at Addis Ababa in 1972. (BAE SYSTEMS)

First VC10 at Heathrow, 1968: G-ARTA was used for extensive testing before conversion to airline standards in 1967 and sold to Freddie Laker, who leased it to Middle East Airlines as OD-AFA. (Malcolm Nason)

VC10 XR806 was the first C Mk1 delivered to No. 10 Squadron RAF in 1966. Shortly after the VC10s had been fitted with refuelling probes, XR806 set two records in 1987, flying from Brize Norton to the Falklands and again back non-stop. Converted to a C1k tanker/transport in 1995 but, two years later, was damaged beyond repair. (BAE SYSTEMS)

After a succession of commercial owners, this aircraft was taken over by the RAF as XX914, its large freight door ideal for service life. Until 1975, it was assigned to the Aero Flight Division as an aircraft for researching structural forces and noise in large jets. Eventually main fuselage sections were used to train loadmasters. (BAE SYSTEMS)

ZA141 was first delivered to BOAC as G-ARVG, it was sold to Gulf-Air as A40-VG, and then acquired by the RAF in 1977 and stored until it became the first K2 conversion in 1982. It was the only VC10 painted in camouflage, which lasted a very short time. After fifteen years in military service the airframe was scrapped November 2000. (BAE SYSTEMS)

A K3 tanker ZA150 streaming a hose from each refuelling point in 1986. Built in 1970 for East African Airways as 5H-MOG, it was repossessed by BAC in 1977 and sold to the RAF a year later. It was the first K3 tanker to fly and, in 1991, the first to fly a Desert Storm tanker mission. It was the first VC10 painted into the grey colour scheme and flew the last operational sortie with ZA147 in September 2013. Now preserved in taxi-able condition by The Brooklands Museum. (BAE SYSTEMS)

The left front door of the K2 and K3 VC10s was converted into an escape chute, which extended into the airflow to enable a safer abandonment of the aircraft. The device was extensively tested but never regarded as reliable and it was never actually used. (BAE SYSTEMS)

Last one on overhaul: the interior of ZA147 Vickers VC10 K3, showing the tanks in the fuselage. This was the last VC10 to fly in September 2013. (Kevin Slade)

Another view of ZA141/B while on finals to Nellis AFB, Nevada, while participating in exercise 'Red Flag' during 1996. The two hose and drogue pods used for refuelling can be seen clearly under the wings. (Rob Finch)

A view of the cockpit of XV106, a RAF VC10 C.1K, on display at Kemble, 20 June 2010. (Kevin Slade)

VC10 K4 ZD240 lands at RAF Waddington in July 2005, a month before being withdrawn from service. (Jonathan Francis)

RAF VC10 K3s ZA147 F and ZA150 J overfly RAF Brize Norton in 2013. (Jonathan Francis)

Bibliography

Andrews, C. F. & Morgan, E. B., *Vickers Aircraft since 1908* (Oxford: Putnam, 1988).

Andrews, C. F. & Morgan, E. B., *Supermarine Aircraft since 1914* (Oxford: Putnam, 1981).

Blackman, T., *Vulcan Test Pilot* (London: Bounty Books, 2015).

Brent, W., *Rhodesian Air Force – the Sanctions Busters* (Nelspruit: Freeworld Publications, 2001).

Brookes, A., *Victor Units of the Cold War* (Oxford: Osprey Publishing, 2011).

Buttler, T., 'Flying the flat iron' in *International Air Power Review* 14 (2004).

Cicalesi, J. C. & Rivas, S. (with Huertas, S. M.), 'Canberras of the Fuerza Aerea Argentina' in *Wings of Fame* 17 (1999).

Darling, K., *Avro Vulcan, Warbird Tech Series* (United States: Speciality Press, 1999).

Donald, D., 'Vickers Valiant' in *International Air Power Review* 25 (2005).

Eckersley, C. F., 'English Electric's Cold War aircraft Canberra, Lightning and beyond' from 47th AIAA Aerospace Sciences Meeting, 5–8 January 2009, Orlando, Florida.

Ellis, G., 'Brothers at arms – Javelins in Zambia' in *The Aviation Historian* (May 2016).

Gardner, B., 'From Provost to Scimitar' in *Aeroplane* (July 2007).

Hamence, M. & Brent W., *Canberra in Southern Africa Service* (Nelspruit: Freeworld Publications, 1998).

James, D. N. *Gloster Aircraft Company* (Stroud: Tempus, 1999).

Lamb, L., 'RAF Operations in Borneo' in *The Proceedings of the Royal Air Force Historical Society* Issue No. 13, Royal Air Force Historical Society (1994).

Lee, D., 'The Royal Air Force Contribution' in *The Proceedings of the Royal Air Force Historical Society* Issue No. 13, Royal Air Force Historical Society (1994).

Mason, F. K., *Hawker Aircraft since 1920* (Oxford: Putnam, 1993).

Manning, C., *Fly Navy: The View from the Cockpit 1945–2000* (Barnsley: Leo Cooper, 2000).

Newark Air Museum Archive. 'Roland Beamont interview.' http://www.newarkairmuseum.org/ (accessed February 2002). Extracts published with permission.

Royal Air Force Historical Society, 'The Canberra in the RAF' in *RAF Historical Society Journal* 43A (1993).

Skinner, S. (ed.), *VC10 – The Story of a Classic Jet Airliner* (Ramsbury: The Crowood Press Ltd, 2015).

Wrigley, G., *Against All Odds. Confessions from an Extraordinary and Varied Flying Career* (United States: Createspace, 2014).

www.thunder-and-lightnings.co.uk
www.flightglobal.com/pdfarchive
www.jetprovostfile.org